THE MEASUREMENT OF ENVIRONMENTAL FACTORS IN TERRESTRIAL ECOLOGY

THE BRITISH ECOLOGICAL SOCIETY
SYMPOSIUM VOLUME NUMBER EIGHT

The Measurement of Environmental Factors in Terrestrial Ecology

A Symposium of

THE BRITISH ECOLOGICAL SOCIETY

Reading 29–31 March 1967

Edited by

R. M. WADSWORTH

University of Reading

with the assistance of

L. C. CHAPAS
The Grassland Research Institute

A. J. RUTTER
Imperial College of Science and Technology

M. E. SOLOMON
Agricultural Research Council Pest Infestation Laboratory

J. WARREN WILSON
Glasshouse Crops Research Institute

BLACKWELL SCIENTIFIC PUBLICATIONS
OXFORD AND EDINBURGH

SBN 632 04220 6

FIRST PUBLISHED 1968

Printed in Great Britain by
ADLARD AND SON LTD, DORKING
and bound by
THE KEMP HALL BINDERY, OXFORD

CONTENTS

PREFACE

The aim in preparing this volume has been to bring together details of methods which the ecologist can expect to be able to use in the field and to discuss recording and the handling of the records obtained. Many specialist books and symposia dealing with the measurement of the environment have been published in the last few years but the ecologist, because he is not a specialist in the techniques of making these measurements, has often found it difficult to obtain the information on techniques which he needs. That this difficulty is widespread amongst ecologists seems to be confirmed by the large attendance at the symposium. It is hoped that the present volume will help to increase our understanding of the environment of plants and animals by making information on methods more readily available to the ecologist.

The majority of the papers in this volume deal with the measurement of specific environmental factors. These papers mention the recording of the factors which they discuss where this is appropriate. They are followed by four papers devoted to recording. The final paper completes a discussion of data handling which has been started in the preceding papers.

It was emphasized in the papers and the discussions at the symposium that it is now easy to measure and record the environment but that it is more difficult to be sure before the start of an investigation what measurements are needed. For this reason it is tempting to try to record everything, or at least those things that are easy to record, and later to try to sort out which factors are important. The sorting of a mass of data is likely to be laborious and expensive even if recording is done directly onto tape which can be fed into a computer. J.W.Siddorn's paper in particular considers this difficulty and it should be early reading for those who are thinking of measuring the environment of an organism. The two papers which discuss investigations where environmental measurements have been used in the elucidation of specific ecological problems may also be helpful here.

When considering one's own investigations the selection of methods is often quicker and easier after seeing examples of the apparatus available and after discussion with someone who has had experience of measurement. At the symposium this was achieved at the demonstration of apparatus around which there was much discussion. These demonstrations are summarized towards the end of this volume where emphasis has been placed on the practical details of manufacture and use, where this is appropriate. For the reader this is a poor substitute for the demonstration and so the address of each demonstrator and of all those who attended the symposium are printed at the end of this volume in the hope that it may be possible for everybody to find somebody near to them with whom they can discuss their problem.

The subject of this symposium was suggested by Professor A.H.Bunting and the symposium was held at the University of Reading in March 1967. I should like to take this opportunity of thanking those at Reading University who helped so much to make the symposium a success and to mention particularly Margaret Quin for taking over all the booking and Christopher J.Smith for arranging the demonstration. Finally I should like to thank those who helped to get this volume to the publishers by speedily editing the following papers: papers 1 and 9, J.Warren Wilson; papers 4, 7, 11 and 14, M.E.Solomon; papers 8, 12, 13 and 15, A.J. Rutter; papers 17, 19, 20 and 21, L.C.Chapas.

R.M. Wadsworth

INSTRUMENTS AND TECHNIQUES FOR MEASURING THE MICROCLIMATE OF CROPS

IAN F. LONG

Rothamsted Experimental Station, Harpenden, Herts.

SUMMARY A good meteorological station is desirable alongside the experimental site, to measure general local climate. The site should be free from topographical irregularities, and the cropped area should be big enough to provide adequate fetch, and to minimize advective effects. Among quantities to be measured are: temperature (air, soil, plant); humidity (vapour-pressure in the air, water content in the soil, dew and leaf-wetness on plants); ventilation; and radiation (total income and components, net and reflected). All need to be measured at several levels (in air, above and within the canopy; in soil, to beyond the expected range of root growth). Continuous recording is desirable, accepting some uncertainty in periods of rain. Standard equipment is rarely adequate. Ignoring radiometry, the paper gives construction details, and examples of performance of instruments designed to measure temperature, humidity and air-flow.

Temperature Thermocouples, thermistors, and resistance thermometers are compared, with examples of advantageous use. Resistance thermometer construction is described, to produce units adequate for measuring soil temperatures, leaf temperatures, and air temperatures with both dry- and wet-bulbs.

Humidity The wet- and dry-bulb psychrometer is described, with corrections for less than full ventilation, and a chart for converting temperature readings to vapour pressures. A variation in circuit design will give an approximate estimate of dew-point temperatures, and two such units, at two heights, will give, more accurately, the difference in dew-point between the heights. Three such units (near ground, in canopy, above canopy) can reveal times of onset and duration of persistence of dew, and its source. Very light deposits can be detected by a surface wetness recorder: irrigated and non-irrigated crops give different responses.

Soil water content can be measured by a neutron scattering technique that shows where the water is being extracted.

Air-flow Cup anemometers are used above the crop. A switch system (designed to avoid some foreseeable sources of error) gives one impulse for about 60 m run of the wind. Logging is automatic (graphical and digital), either on mains or battery operation. Within the crop, a hot-bulb anemometer takes up very little space and is more sensitive than rotating cups.

INTRODUCTION

The microclimate of field crops is determined by the fluxes of water vapour, heat and momentum between earth, plant and atmosphere. Measurement of the crop environment is concerned mainly with the distribution of air-flow, temperature, humidity and radiation, above and within the crop, and the temperature and moisture content of the soil. Account must also be taken of fluctuations in the energy and water balance, the friction between plant and air, and environmental factors which affect photosynthesis, such as carbon dioxide concentration, and solar radiation in the 0·4–0·7 μ waveband. As these factors are influenced both by the physical and physiological characteristics of the plant, measurements are required of crop growth and development. The most important of these are, height, cover, leaf area, state of crop, and details of response to farming operations or cultivation techniques.

EXPERIMENTAL SITE

A study of the environment of a field crop may be of little value without measurement of the general climate of the locality. A good meteorological station, close to the experimental site, should provide measurements of rainfall, temperature, humidity, wind velocity and direction, and total solar radiation. The records will allow the local macroclimate to be compared with the microclimate of the crop, and may be used to estimate other meteorological factors such as potential evaporation (Penman 1963).

Estimates of the total heat and mass exchange per unit area of a crop can be estimated from measurements of the vertical variation of temperature, humidity and air-flow above the crop (Penman & Long 1960), but these estimates are only representative of the crop when horizontal advection is insignificant. To satisfy this condition there must be sufficient fetch, particularly upwind: it is generally agreed that the fetch should be at least 100 times the interval between the highest measuring point and the top of the crop canopy.

The experimental plot should be uniform in height and homogeneous in character, and the area surrounding it should be covered by crops preferably of similar height and cover. When there are large differences in cover, or the site is surrounded by fallow areas, horizontal advection may be serious, resulting in the so-called 'oasis' effect (Brooks 1959) This advection can greatly distort comparisons of microclimate made between different crops or crop treatments, particularly when the plot areas are too small. When comparing irrigated with unirrigated crops, large plot sizes are especially essential. Water will evaporate much faster from small wet plots surrounded by large relatively dry areas, than from similar plots in the middle of a large irrigated area. Choice of an ideal site is not always possible, but a preliminary survey of the variation of meteorological factors over the entire experimental area may reveal a best site.

The potentials and their gradients, measured at a chosen site, depend on the surface conditions upwind. They are representative of the site only when the surface conditions upwind are the same as those of the site. With sufficient fetch, a short crop or bare soil meets these requirements, but for tall crops, such as cereals, some distortion in vertical profiles cannot be avoided. The distortion may be minimized by averaging the vertical profiles of several measuring sites placed at regular intervals within the experimental plot.

MEASUREMENTS

All weather elements interact one with the other, so that an adequate knowledge of the environment of a growing crop requires the long term and continuous recording of many factors. To measure only one is rarely useful.

Because the environment of a crop is three dimensional, interpretation is required of the spatial distribution of the climatic factors within and above the crop canopy. Unfortunately expense, labour and shortage of equipment usually restrict the measurement to a study of vertical profiles at a site assumed to be representative of the whole area.

The plant itself, as part of the system, contributes to its own climate, and to that of its neighbours. Its complex growth, geometry and needs change both with development and age: these factors should be observed and, where possible, measured.

All the physical potentials associated with the crop are determined by the incoming solar radiation and by the available water in the soil. An analysis of the radiation climate requires measurement of the total incoming solar radiation and the component parts of the resultant energy fluxes. (The many types of radiation measuring instruments are described and their performance compared by Robinson 1966).

The absorption of carbon dioxide by the plant is also part of the microclimatology, and measurement is required of its fluxes and concentrations within and above the crop, and in the soil. Because the photo-chemistry of assimilation occurs only in a limited part of the visible spectrum, knowledge is also required of the intensity and distribution with height of this fraction within the plant canopy. This can be attained by profile measurements of total radiation inside the crop, and suitable filtering of an identical set of radiometers at the same levels. (Solarimeters and albedometers suitable for use in and above crops are described by Monteith & Szeicz 1962, and by Szeicz, Monteith & Dos Santos 1964). Account must also be taken of the pattern of sunlight interception and the air flow characteristics of the vegetation canopy, both of which affect the direct exchange of carbon dioxide between plant and air.

Instantaneous rates of energy transfer of a crop can be calculated from profile measurements of temperature, humidity and air-flow, but very great accuracy is needed both in the measurement of the potentials and in the placement of the sensors (McIlroy 1955). Such accuracy can be obtained from very delicate instruments used over short periods of time, but is almost impossible over long periods, and for biological and microclimate studies that extend over the whole growing period of a crop, less precise measurements must be accepted. Long-period, continuous records, although less accurate, may, with reasonable care and some experience, be used for rough calculations of the energy and water vapour fluxes, provided the data are averaged over periods of an hour or more (Penman & Long 1960). This integration is required because all meteorological factors measured near the ground show random fluctuations, with amplitudes and periods determined by the strength of turbulence and the vertical and horizontal distribution of the meteorological factors. The periods are usually not more than a few minutes, and can be neglected when records are integrated over a long enough period. However—as a warning—sometimes there are fluctuations with a period equal to the printing period of the recording system, and then even integration over a period of one hour will not eliminate distortion.

The short-period fluctuations should not be confused with the longer-period changes, i.e. diurnal variations and those caused by air-mass advection: or, with the medium period fluctuations in local advection arising from cloud cover, gusts of wind, and inclination and direction of the sun.

Few meteorological instruments function correctly when it is raining, a problem that still remains unsolved. In heavy rain, cup anemometers may rotate backwards, and, through insufficient screening, dry-bulb thermometers may become wet. For the same reason, hot-bulb anemometers and radiometers will not function correctly. This is particularly true of the small sensing units used in measuring the microclimate. Micro-meterology tends unfortunately to be a 'fine-weather' subject, with the inevitable consequence that choice of periods for analysis has to be selective. For this reason, and partly because the weather is still largely unpredictable, long-term, continuous recording is essential. A further justification of continuous recording is that the equipment is in use whenever an event of interest occurs. For example, knowledge may be required of the occurrence of dew in the days preceding an outbreak of fungus disease in the crop. The profile records will show the start of condensation, how long it lasted, its approximate intensity, and how long it took to evaporate. Such records, supplemented by the direct measurements of a mechanical 'dew-balance' or those from a 'surface-wetness' recorder can be very useful to plant pathologists.

The meteorological potentials of temperature, humidity and air-flow above a surface follow an approximate logarithmic distribution with height, and thus the choice of a logarithmic scale for the measuring heights of the instruments is advantageous. However, above a crop or rough surface, the logarithmic distribution does not start at the ground surface, but from a higher level. Over bare soil this 'zero plane displacement' is small and constant. With a crop cover it is at about two-thirds of the crop height, but varies considerably with the velocity and turbulence of the air, with the physical characteristics of the plant, and with the surface wetness of the leaves. Within the crop, the instruments are best spaced evenly with height. Analysis of the records and assessment of the zero plane is greatly facilitated by fixing the profile instruments above the crop to a telescopic mast able to 'grow' with the crop. At least five measuring heights are required above the crop, and five inside.

INSTRUMENTATION

Measurement of the plant environment usually necessitates some modification of standard equipment or the design of special sensors for use in small spaces. The choice and design of sensors are governed by the purpose of the measurements, the accuracy required, the recording system available, and by the cost.

The sensors and accessories should not interfere with the growth of the crop or with the natural meteorological processes taking place. For example; a large sensor and its mounting system may act as a heat sink or as a disturbing source of radiation: large radiation screens can hinder radiation exchange within the crop, particularly in dense vegetation: and ventilated sensors, by intermixing the surrounding air, may destroy the natural potential gradients (Franssila 1936). The merits of the various sensors available are fully discussed in the literature, (e.g. Tanner 1963). The many recommendations may at times appear contradictory, but the explanation usually lies in the different circumstances in which the authors have applied the sensors. Familiarity with a particular technique tends to produce a bias. The following comments on sensors and instrument design are based on experience gained in measuring the physical environment of growing crops during the past twenty summers at Rothamsted; they too may contain bias. The discussion will be limited to the measurement of temperature, moisture and air-flow.

1 *Temperature*

There are in popular use three main types of sensor suitable for remote indication and recording of temperature within the microclimate range, −25°C–50°C. They are, the wire type resistance thermometer, the thermistor, and the thermocouple. Each has its own particular advantages and disadvantages. Their major characteristics are compared in Table 1.

(i) *Resistance thermometers*

The electrical resistance of most pure metals increases with the temperature and this property is exploited in the wire resistance thermometer. For platinum the rate is about 0·4 per cent and for nickel about 0·6 per cent per degree centigrade. Over the temperature range −25°C–50°C the platinum sensor departs from linearity by about 0·1 per cent and the nickel sensor by about 0·2 per cent but provided identical sensors are

TABLE 1. Comparison of temperature sensors.

Characteristic	Wire sensor	Thermistor	Thermocouple
Linearity	Excellent	Non-linear	Excellent
Sensitivity	High	Medium	Low
Stability	Excellent	Poor	Good
Self heating	Low	High	None
Interchangeability	Excellent	Poor	Good
Spatial sensing	Good	Poor	Poor
Point sensing	Poor	Good	Excellent
Difference sensing	Good	Poor	Excellent
Cost	Dear	Moderate	Cheap
Recorder cost	Moderate	Moderate	Dear

used, this slight deviation from linearity may be neglected in field work.

Because of their high stability, wire sensors are best used for measuring absolute temperatures. When properly constructed and annealed, the small platinum sensors used in micrometeorology are stable to within 0·01°C per year. They are equally well suited for the measurement of temperature differences, provided care is taken with lead compensation, and frequent checks are made on possible recorder drift. A particular advantage of wire sensors is that, because they are linear, several units may be connected in series to provide a spatial average of temperature. They can also be wound to small dimensions and used for point or surface measurements, but for these purposes thermistors or thermocouples are much better.

The resistance recorders used with wire sensors must be very stable, have a linear response and be free from drift. The type generally employed is the double-slidewire version of the Wheatstone bridge, with a null balance system, developed by M.E.Leeds over 50 years ago. (For a full description see Middleton & Spilhaus 1953).

Suitable miniature wire sensors are obtainable in various forms. The most stable and robust are the platinum types wound and embedded in borosilicate glass. A typical unit has a nominal resistance of 100 ohms at 0°C and a sensitivity of 0.385 ohm/°C. Dimensions are 3 mm diam and 20 mm length. At Rothamsted we encase them in brass heads, using 'Araldite' as the embedding medium, and mount them at the end of pyrex tubing 3 mm diam, length 65 mm. The 'Araldite' gives a water-tight seal that has withstood more than two months continuous immersion in water during calibration and testing operations. Final dimensions of the sensing heads are 4 mm diam, length 22 mm. In still air (velocity less than 5 cm/sec) the time required for a temperature difference between thermometer and air to fall to 1/e of its initial value (Platt & Griffiths 1964) is 90 sec when used as a dry-bulb thermometer, and 70 sec when used as a wet-bulb thermometer.

For special purposes, or when it is necessary to match sensors to an existing recorder, it is better to construct the thermometers. We have used enamelled nickel-wire, 45 s.w.g., non-inductively wound on perspex bobbins (diam 3 mm, core diam 1 mm, length 10 mm; Fig. 1a). They are encased and mounted like the platinum units, and have similar dimensions and thermal lag. Typical units built have a resistance of 68 ohms at 0°C, and a sensitivity of 0·36 ohm/°C. No change in characteristics has been detected over a period of 10 years (Long 1957).

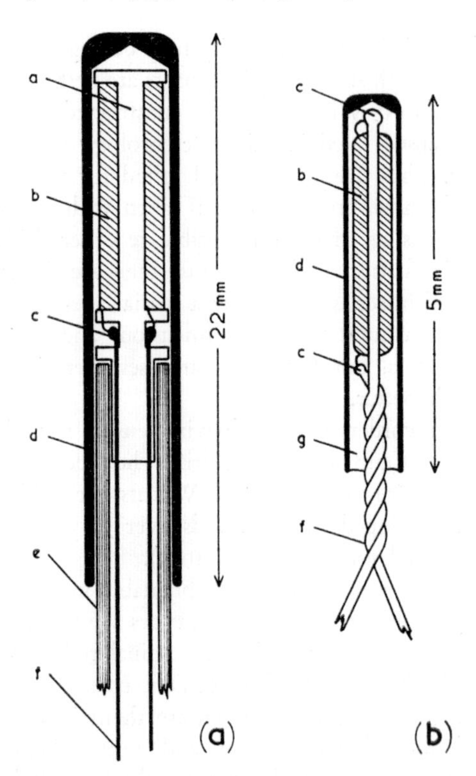

FIG. 1. Construction of standard nickel wire thermometer, a; and leaf thermometer, b. (a) perspex bobbin, (b) nickel wire coil, (c) soldered joint, (d) brass sheath, (e) glass tube, (f) copper leads, (g) thermal setting resin.

Similar nickel sensors mounted at the end of pliable polythene tubing (100 cm long) are suitable for measuring soil temperatures. They can easily be inserted into holes bored into the exposed face of a pit dug in the soil. Eleven such units have been used to give a continuous record of soil temperatures to a depth of 1 m under grass. They remained buried for 3 years, and during this period no failures or drift in characteristics were detected.

Very small nickel sensors have also been constructed to measure leaf and plant temperatures. With some difficulty it has been possible to wind 50 s.w.g. enamelled wire onto a copper former (38 s.w.g. enamelled), and to machine a metal sheath, so that the external dimensions of the sensing element are 0·8 mm dia. and 5 mm long (Fig. 1b). As in the larger units, the nickel resistance coil was embedded in a thermal setting resin within the metal sheath. The units are small enough to be inserted into the thicker parts of sugar beet, kale or potato leaves, after making a hole with a number 18 hypodermic needle. On grass or cereal leaves they are wrapped and tied with cotton in the folded leaf. They can be stitched onto the leaves of field beans. These sensors are also used as soil surface thermometers.

The sensitivity of these thermometers is the same as that of the larger units. In still air the time-lag constant is 9 sec.

(ii) *Thermistors*

Thermistors are solid-state semi-conductors which exhibit large non-linear changes of resistance with temperature. They are composed of sintered mixtures of very pure metallic oxides, such as nickel, manganese, copper, iron and uranium. They are available as beads, rods or discs in various mountings and lead configurations. Their resistance may increase or decrease with temperature by from 3 to 6 per cent/°C, dependent upon the oxide constituents. Most of those commercially available have a negative temperature coefficient with a near logarithmic response in the range −25°C–+50°C.

Thermistor beads, with diameters of 1 mm and smaller, respond very rapidly to changes of temperature, but their thermal capacity can be increased by encapsulation in a suitable container. Thermistors are best used for temperature measurement at a point, and are especially useful for measuring plant temperatures. They are easily attached to a leaf surface, or inserted into plant material.

The high temperature coefficient and small energizing currents required, make the thermistor suitable for use with portable, battery operated, Wheatstone bridge circuits. A portable, robust and inexpensive thermistor bridge for psychrometric and air-flow measurements in crops has been in use at Rothamsted for 20 years (Penman & Long 1949). It has proved reliable for spot measurements, so long as frequent checks are made on calibration.

In many applications the non-linearity of thermistors is a serious disadvantage, particularly if they are used for measuring temperature differences. In simple theory linearization cannot be achieved, but in practice it can be closely approximated over a small range, by a combination of series-parallel resistors. At Rothamsted we have used this technique to linearize thermistors in the range 0–50°C and matched them to a linear resistance recorder (Kent 12 point Multelec) with an 18 ohm range. The thermistors used have a resistance of about 5000 ohms at 0°C, 2000 ohms at 20°C, and 700 ohms at 50°C. A resistor, R_s, is put in series with the thermistor and another, R_p, in parallel with the combination. Values are:

$$R_p=\left[\frac{R(r_t+r)\,(r_0+r)}{r_0-r_t}\right]^{1/2}\simeq 206 \text{ ohms}$$

and

$$R_s=r-R_p \qquad \simeq 794 \text{ ohms}$$

where R_p = parallel resistor
R_s = series resistor
R = 18 ohms for Multelec recorder
r = 1000 ohms for thermistor type F2311/300
r_0 = resistance of thermistor at 0°C
r_t = resistance of thermistor at 50°C.

Experience shows that, within the range 0–50°C, nearly all types of thermistor may be linearized by this method, so long as the resultant resistance change, R, from 0–50°C, is not greater than $0{\cdot}02r$. The value of r determines the degree of linearity, and is best found by trial: it is approximately $0{\cdot}25\,(r_0-r_t)$. When recording at a distance, the resistance of the field leads is included in the resistance R_s. It is not necessary to employ temperature compensation for the field leads, because the temperature coefficient of the leads is negligible compared with that of the thermistor.

Most resistance recorders are designed for use with positive temperature coefficient sensors and give a left to right deflection with increasing temperature. The same deflection can be obtained with thermistors by placing them, and their resistance networks, in the opposite arm of the bridge circuit to normal. Some adjustment of the recorder zero may also be required.

Thermistors are not directly interchangeable, because both temperature coefficient and initial resistance vary from unit to unit, so that each individual thermistor requires calibration and trimming of linearizing networks. At Rothamsted, we used linearized thermistors to record the environment of a potato crop during two seasons (Hirst, Long & Penman 1954). Frequent spot checks with an Assmann psychrometer indicated that mean air temperatures were correct to 0·05°C and derived dew-points to within 0·1°C.

In general thermistors are not suitable for continuous recording. We found that the long term instability, the frequent necessity for recalibration, and the corrections required make analysis of the data very laborious, and since 1954 we have used wire sensors. In recent years performance of thermistors has improved considerably. Of those available, the hermetically sealed, glass-bead thermistor has the best long-term stability, but variations still occur which have no predictable pattern and they do not improve with ageing.

(iii) *Thermocouples*

When dissimilar metals are joined together an electric potential difference is produced at the contacting surfaces. When two junctions are formed by connection two dissimilar metals in a loop, then a current will flow in the circuit, proportional to the net potential resulting from any temperature difference at the opposing junctions.

The thermocouple detects temperature difference, and to measure the ambient temperature, one junction must be held at a known constant reference temperature. The ambient temperature is then the sum of the reference temperature, and the temperature difference detected between reference and measuring junction. Slowly melting ice, in distilled water contained in a thermos flask, gives a stable temperature reference for field work, provided the ambient temperature stays above 0°C, and the reference junction is not in actual contact with the ice. In cold weather the reference junction may be buried in the soil at a depth of 2 m, where the temperature rarely varies by more than 0·05°C in a year.

Provided it does not vary too greatly the ambient temperature, t_a, at the recorder may also be used as a reference junction. Variations in t_a are electrically compensated by means of a series-connected Wheatstone bridge circuit in which a wire sensor, in thermal contact with the reference junction, acts as the temperature standard.

In the bridge network (Fig. 2), R_1 equals R_2, and R_3 equals the resistance of the wire sensor, W at 0°C. With both junctions of the thermocouple held at mean t_a, the bridge is balanced for zero voltage between A and B, by the potentiometer R_4. When the reference junction temperature varies from mean t_a, the wire sensor unbalances the bridge and a voltage appears between A and B, in opposition to, and compensating for, the change in potential at the reference junction. The level of the compensating voltage between A and B is preset by R_5 to equal the output voltage

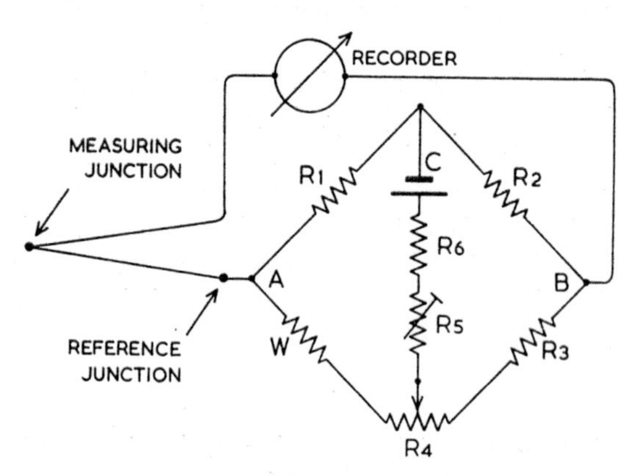

FIG. 2. Electrically compensated reference junction. R_1, R_2, 50 ohms; R_3, 100 ohms; R_4, 30 ohms; R_5, 5 k ohms; R_6, 2·5 k ohms. W, Platinum wire sensor, 100 ohms at 0°C. C, mercury cell 1·4 V.

of the thermocouple. One advantage of the system is that the recorder zero may be offset by introducing a voltage, negative or positive, into the thermocouple loop by adjustment of R_4.

The response of a thermocouple is almost linear, but the output is very small—around 40 μV/°C—so that a sensitive amplifier is required to actuate a recorder. By contrast, the output from a wire-sensor-bridge would be at least 4 mV/°C, sufficient to drive a recorder without amplification.

Thermocouples are not the best choice for measuring ambient temperature in the field. In addition to the inconvenience of maintaining a reference junction, careful screening of long leads is required. Alternating current pickup is partly rectified at the thermojunctions and may mask weak signals. Also, when large ranges of temperature are covered, the current flowing in the thermocouple circuit tends to heat one junction and cool the other, thus upsetting the accuracy of the reference junction.

Both these effects can be partly decreased by increasing the impedance of the thermocouple circuits and by using a high impedance potentiometric recorder.

Thermocouples are best used for measuring temperature differences, and when made of very fine wire, so that the junctions have both small size and thermal capacity, they are very suitable for measuring differences in temperature between leaves and the air. One junction is placed in the leaf and the other at a fixed distance from the leaf surface. Several sets of thermocouples, wired in series, will give a spatial average of the leaf-air temperature difference.

Thermocouples are inexpensive and easy to make. Copper and constantan are good choices for construction because the materials are relatively homogeneous, and elements made from the same batch of wire are interchangeable. However, although the sensor cost is small, the recording systems, compared with wire-sensors or thermistors, are relatively expensive.

(iv) *Temperature errors*

The temperature indicated by a thermometer is a summation of convective heat exchange with the air flowing past it, and of radiative exchange with nearby surfaces, ground, sky and sun. The true air temperature is measured only when thermometer and air are in thermal equilibrium. Temperature errors arising from direct radiation can be diminished by screening the sensor, and by employing highly reflective coatings on both sensor and screen. A suitable coating is a matt-white paint that 'weathers' so that its surface is renewed. A surface of 'Mylar'-based aluminium tape is also very efficient. (For a survey of reflective coatings, see Tanner 1963). Radiative exchange can be further decreased and convective exchange increased, by using small sensors: however, the increased speed of response may be undesirable when temperature integration is required over a period. In addition the response time should not be significantly smaller than the printing period of the recorder system. The final size of the sensor must be a compromise between an acceptable radiation error and the time constant required. Convective exchange can be increased by forced ventilation of the sensor, but this is not recommended in profile studies, especially within the crop canopy.

The self heating of wire sensors and thermistors will also introduce temperature errors when too large an energizing current is employed in the measuring circuits. The thermistor, because of its small size, is particularly prone to this error, because the heat generated must be dissipated through a small surface area. For example; a dissipation of 1 mW will increase the temperature of a bead thermistor, which has been encased in a heat sink, by about 1 °C. By contrast, the surface area of a wire sensor in thermal contact within a similar heat sink would be very much greater, and would require a dissipation of at least 20 mW to raise its temperature 1 °C.

2 *Humidity*

The measurement of humidity in growing crops is more difficult and less accurate than that of temperature and many types of sensor have been developed for humidity measurement. (For descriptions, theory, applications and many references, see Wexler 1965). None is completely reliable, and few can be recommended for continuous recording in the field. The most accurate system suitable for continuous recording is the infrared hygrometer. It measures the absolute humidity, and an accuracy of around $\pm 0{\cdot}0025$ g/m^3 can be expected (Brooks 1963). If the relative humidity or vapour-pressure gradients are required, then a complementary system of temperature measurement is necessary.

The system generally used in field work is the wet- and dry-bulb psychrometer: although less accurate, it is adequate for most microclimate studies and is comparatively inexpensive. Discussion here will be limited to details of construction and accuracy of psychrometer units for use in and above the crops.

The wire sensors, described earlier, were mounted in pairs, parallel and 15 mm apart, on an ebonite base (28 mm dia., 10 mm thick). The base fits into a short length of brass tubing (25 mm length), which is closed at its end by a disc of ebonite (4 mm thick). Field leads and sensors terminate inside the base, which is filled with 'Henley's' plastic compound to exclude water. The base has a clip on each side; one for attachment to a post, and the other to hold a reservoir for the wet-bulb. The radiation screen is a hemicylinder of copper gauze, 35 mm dia., and 40 mm long. It is supported and attached to the base by 16 s.w.g. copper wire. The entire unit is painted matt-white (Plate 1).

One sensor has a thin close fitting muslin sheath, 26 mm long, which covers the sensor and 4 mm of the glass support. Four threads of cotton wick lead from the underside of the muslin to a J-shaped glass reservoir mounted alongside. The wick is long enough to pass right round the bend, a condition essential for maintained flow. Experience indicates that the length of wick exposed between sensor and reservoir should be

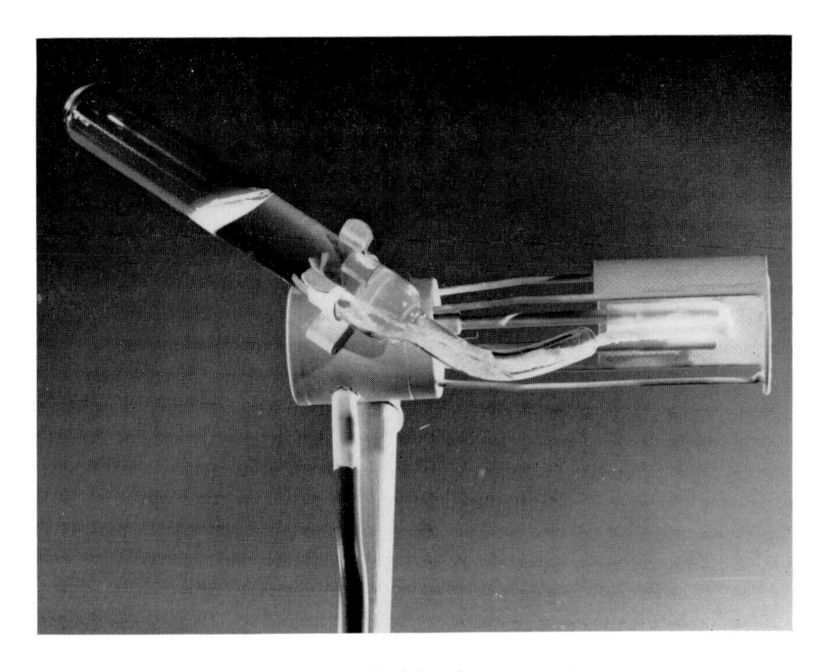

PLATE 1. Standard Psychrometer Unit.

[*facing p.* 14

constant. A standard gap of 5 mm is used. The reservoirs hold 15 cm^3 water, sufficient for one day's supply in dry summer weather. In winter polythene reservoirs are employed, to avoid breakages from freezing.

The psychrometers are mounted on masts and set to face south, with a north–south alignment, so that low morning or evening sun does not shine directly on the sensors. Calibration in full sunlight against an Assmann psychrometer indicates that the radiation shield is adequate, except when the ventilation is slow, or the units are used over very reflective surfaces. In slow ventilation (< 1 m/sec) and full sunshine (1·2 cal/cm^2/min) the dry-bulb temperature may rise to a maximum of 0·1°C above ambient temperature, but the corresponding rise of the wet-bulb temperature is only 0·05°C, with the result that the derived vapour-pressure is almost unaffected.

The use of a screen below the sensors not only increased rather than decreased the radiation error, but also diminished ventilation of the system. Many other designs of screen have been tried, but none was significantly better in anti-radiation properties than the type in current use.

A determination of the psychrometric constant at different wind velocities, made in a wind-tunnel under conditions of laminar air-flow, gave values indicated by the solid line in Fig. 3. The 'constant' decreases rapidly with increasing ventilation, and attains a steady value of 0·67 mb/°C when the velocity exceeds 120 cm/sec. Calibration in the field against an Assmann psychrometer, gave much smaller values, attaining a steady value at 100 cm/sec, (dotted line Fig. 3), indicating that natural turbulent air-flow provides better ventilation of the wet-bulb than laminar air-flow.

3 *Psychrometer chart*

A psychrometric chart has been prepared from which the vapour pressure can be determined from wet-bulb temperature and wet-bulb depression (upper section Fig. 4). The wet-bulb temperature is entered horizontally via the ordinate, and the wet-bulb depression vertically through the abscissa. The intersection gives the vapour pressure, which is read from the diagonal curves. The chart was computed from the standard psychrometric equation:

$$e = e_s - \gamma (T_a - T_w)$$

where e is ambient vapour pressure at T_a—the dry-bulb temperature, e_s is the saturated vapour pressure at T_w—the wet-bulb temperature, and γ is the psychrometric constant (Fig. 3).

The psychrometric constant $\gamma = aP$, where a is a constant dependant upon the physical units employed, and on the geometry and ventilation of the psychrometer, and P is the atmospheric pressure.

The chart is for use with fully aspirated psychrometers at a standard atmospheric pressure of 1013 mb, and a psychrometer constant a of $0{\cdot}66 \times 10^{-3}$ mb/°c. The normal variations of atmospheric pressure from

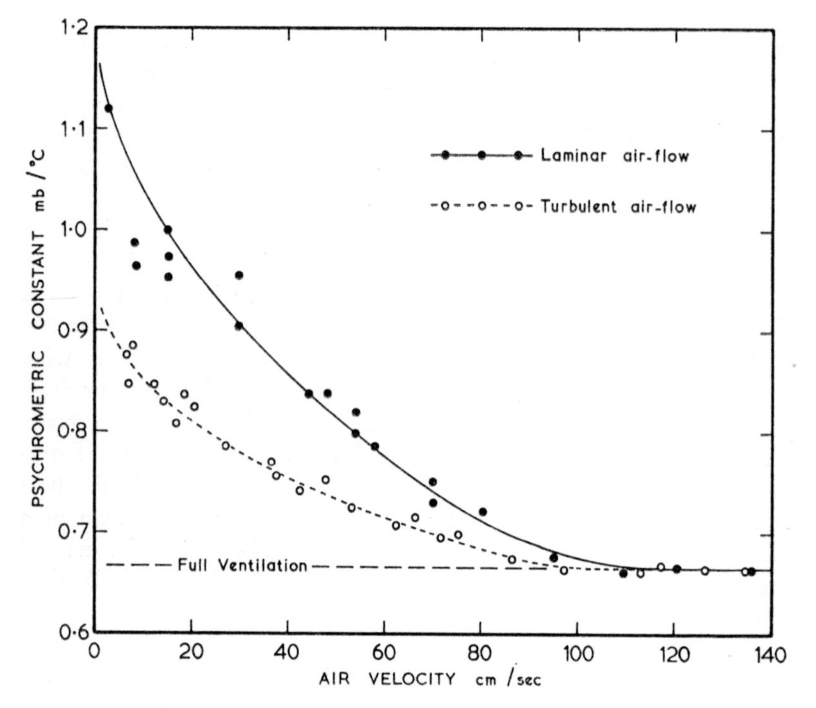

FIG. 3. Psychrometer constant of standard nickel and platinum thermometer units at low air-flow.

1013 mb will introduce small errors in the derived values of absolute vapour pressure, but they are insignificant and may be neglected in micro-climate studies.

A much larger error is introduced when insufficient ventilation of the psychrometer limits evaporation and prevents maximum cooling of the wet-bulb. However, if the ventilation velocity is known, then the wet-bulb depression may be corrected by use of the appropriate psychrometric constant, γ. On the chart (upper section Fig. 4) this is equivalent to expanding the wet-bulb depression scale by the factor $\gamma/0{\cdot}67$. This can

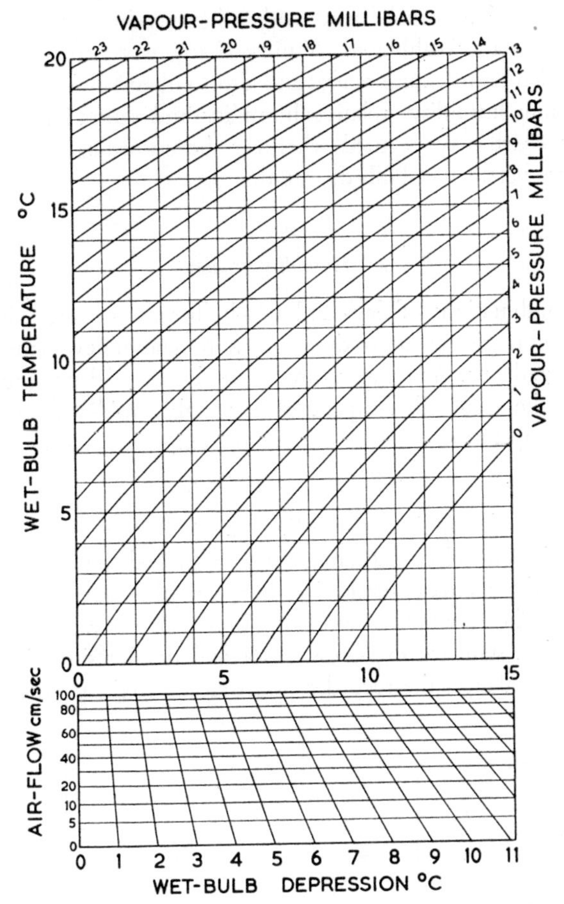

FIG. 4. Psychometric chart: vapour-pressure (upper section); correction for low-airflow (lower section).

be achieved quite simply by using a transparent grid of wet-bulb depression scales, corrected for γ (lower section Fig. 4). In use, the required scale is selected by entering the ordinate at the appropriate ventilation velocity (or correlative γ). The corrected wet-bulb depression scale selected is then superimposed upon the vapour-pressure chart at the relevant wet-bulb temperature. The vapour-pressure is then read from the chart as before, but using the corrected scale.

The values of γ used in the ventilation correction chart are the lower values from Fig 3, and apply only to the psychrometers described here.

However, the chart can be used with other psychrometers, when they have been suitably calibrated. Similar charts were computed for use over extended temperature ranges and also to obtain dew-point temperature, absolute humidity, relative humidity and saturation deficit. The correction chart may be used wherever the co-ordinates are wet-bulb temperature versus wet-bulb depression.

4 *Humidity gradient*

Suitable combinations of dry- and wet-bulb sensors were wired into a bridge circuit to give an approximate measurement of dew-point temperatures and gradients. At temperatures around 10°C and not too far from saturation, the wet-bulb temperature is roughly mid-way between the dry-bulb temperature and the dew-point temperature, i.e.:

$$T_w \simeq 0{\cdot}5\,(T_a + T_d),$$

or in terms of dew-point, $Td \simeq 2T_w - T_a$.

The equivalent electrical circuit was obtained by making sensors with twice the standard sensitivity, but with the same external sheath dimensions· They are used as wet-bulbs in one arm of a recording bridge circuit, with a standard dry-bulb sensor and fixed resistor in the other (Fig. 5a). At balance, the recorder prints an approximate reading of T_d

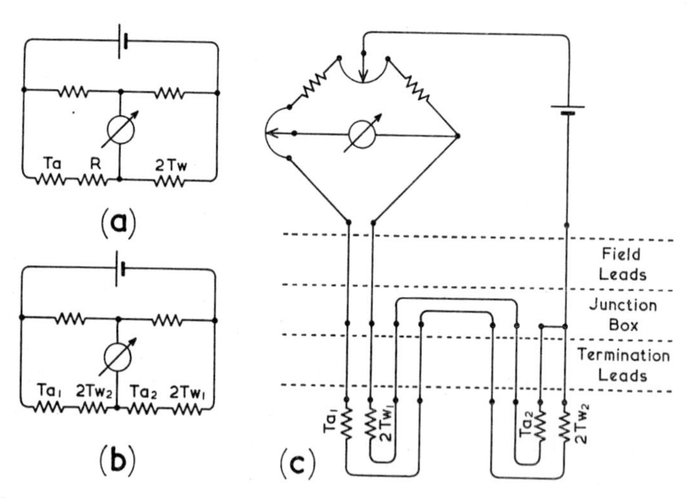

FIG. 5. (a) *Td* circuit. (b) Δ*Td* circuit. (c) Practical circuit of Δ*Td* recorder. For details see text.

directly on the chart. This gives a picture, admittedly somewhat crude, of the daily trend of absolute humidity at the observational level.

A more accurate indication is obtained by direct measurement of the T_d gradient. A pair of T_d units are placed at two levels, 1 and 2, and wired into a bridge circuit so that their deflections are in opposition (Fig. 5b). The recorder deflection is $T_{d1}-T_{d2}$, where:

$$T_{d1}-T_{d2}=(2T_{w1}-T_{a2})-(2T_{w2}-T_{a1})=\Delta T_d$$

The record gives the direction of vapour transfer at any time of day, and can be read at a glance from a fixed reference line on the chart.

Sensitivity is such that ΔTd can be measured with ease to 0·1°C and the sign detected to better than 0·05°C. The error in measured ΔTd increases both with increasing temperature gradient, and with decreasing relative humidity. For example; in the range 5–10°C with a difference of 10°C in Ta over a height interval of 50 cm, ΔTd is underestimated by about 10 per cent at 75 per cent relative humidity; but at 95 per cent relative humidity the underestimate in ΔTd is less than 2 per cent. The change of wind speed with height also introduces a smaller error, but of opposite sign, when the air speed is less than 100 cm/sec. The example given is an invented extreme: real air temperature gradients are very much smaller, and the errors in ΔTd are much less.

The circuit of the ΔTd recorder (Fig. 5c) has lead compensation between the sensors at two levels, and between the sensors and recorder. The ΔTd recorder has been used for 12 years to measure the incidence and ending of dew in various crops. Two gradients are measured, one from above the crop to just below the crop surface, and the other from crop surface to crop interior. The two gradients indicate approximately the relative magnitudes of dew coming from above the crop and dew formed by distillation from the ground. The records, when combined with air-flow gradients, (recorded on another instrument) also enable the total dewfall and its intensity to be estimated approximately.

The apparatus is simple to instal and has negligible effect on the local environment. The traces characteristic of dew formation are easily recognized, and comparison with records from a mechanical 'dew balance' (Hirst 1954) and a 'surface-wetness' recorder, indicate that the times of onset and finish of dew deposition can be determined by inspection to within 5 min. The conditions for dew formation are clearly recorded, and are not affected by the occurrence of rain before or after deposition, or by a surface already wet. Guttation, easily mistaken for dew, (Long 1958) is not recorded. Short periods of dew deposition are

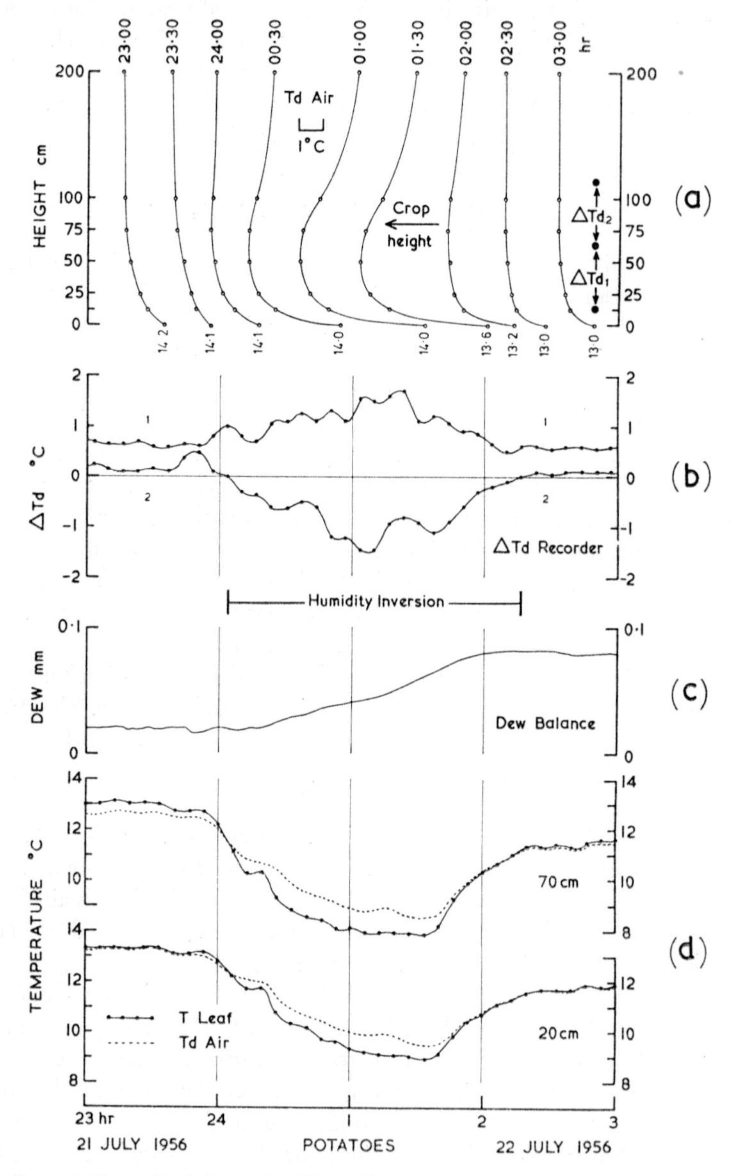

FIG. 6. Record of dew. (a) *Td* profiles. (b) ΔTd record. (c) Dew-balance record. (d) leaf temperatures.

detectable and on several occasions dew formation lasting less than 15 min has been recorded. Periods of extremely light dew, which are not always revealed by a mechanical balance, are also detected, and this has been confirmed by comparison with an 'electronic surface-wetness' recorder (Section 5 p. 21).

A typical record of dew (Fig. 6b) is for a potato crop at Rothamsted during the night of 21 to 22 July 1956. Trace 1 indicates the gradient within the crop, and trace 2 the gradient above the crop. When the traces are positive water vapour is moving from the lower to upper level (i.e. evaporating), and downward when negative. The night was mainly overcast, but a gap in the clouds from 23·50 to 01·40 hr resulted in an increase of outgoing radiation and the potato leaves cooled rapidly to about 1°C below the ambient dew-point temperature. The leaf temperatures and the dew-point temperature of the air at 20 and 70 cm heights are shown in Fig. 6d. Traces 1 and 2 (Fig. 6b) show the sudden change in vapour pressure gradients and the resultant humidity inversion, clearly indicating that conditions for dew formation existed between 00·10 and 02·10 hr. The corresponding record of dew from the mechanical dew-balance is on Fig. 6c. For further comparison, the dew-point temperature profiles, in and above the crop, were determined from the records of T_a and T_w (Fig. 6a). The pause in the cooling rate of the potato leaves, occurring just after the start of dew deposition at 00·10 hr, is probably a latent heat effect; many similar examples occur in the leaf temperature records. Average air-flow during the 4 hr period was 59 cm/sec at 2 m, 31 cm/sec at 1 m and < 15 cm/sec in the crop.

5 *Surface-wetness recorder*

The presence and persistence of surface water films on leaves is an important factor both in the plant's environment and in the development and spread of some plant diseases: many kinds of fungus spores require a continuous film of liquid water at certain stages of their growth, and to facilitate their physical movement over plant surfaces. The water films may originate from rainfall, mist, fog, dew or guttation (Long 1955, 1958), events which are usually recorded or can be deduced from standard meteorological observations. The ΔTd recorder reliably indicates the period of dew precipitation, and the mechanical dew-balance gives the total surface water on the plant from all sources, and its persistence. Neither instrument, however, provides a profile measurement of the persistence of surface water films within the crop, and for this purpose an electronic surface-wetness recorder has been built. Suitable pairs of

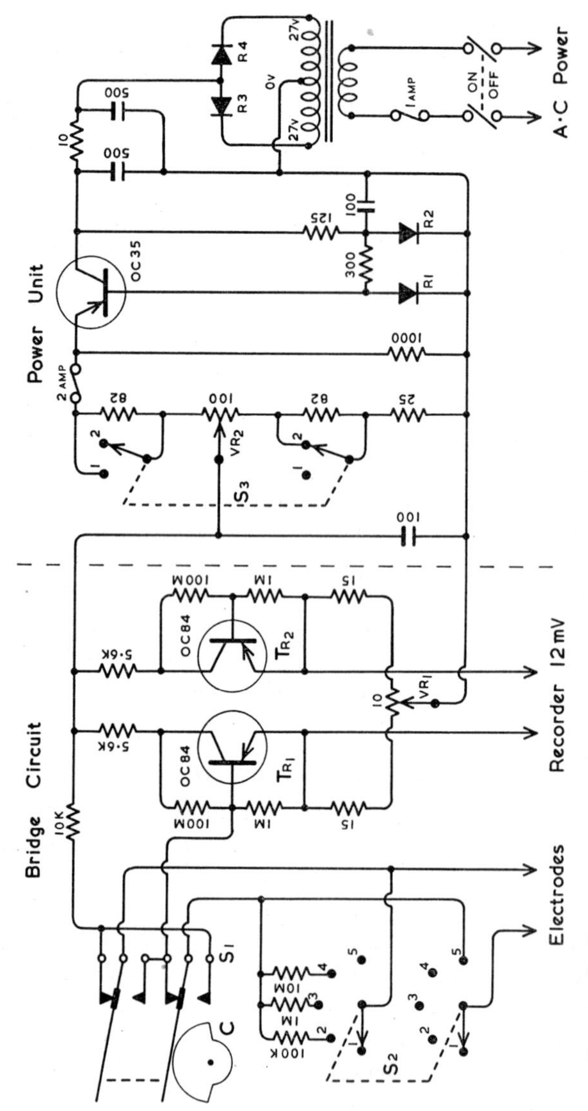

FIG. 7. Surface wetness recorder. All circuit values in ohms or μF unless specified. R_1, BZY95-C24; R_2, BZY95-C16; R_3, R_4, Rec 30.

electrodes are clipped on the leaves at various heights in the crop, and the changes in electrical resistivity resulting from the surface water films are measured by a transistor bridge circuit (Fig. 7).

The bridge circuit consists of two germanium transistors Tr_1, Tr_2, and their respective resistor networks. The transistors are matched to give temperature compensation, and the resistor networks provide negative feed-back to stablize the circuit. For maximum stability Tr_1 and Tr_2 are kept at constant temperature. The electrodes are clipped to a leaf and permit a small base current to flow in Tr_1. The current is inversely proportional to the electrical resistance of the leaf, and increases with increasing leaf wetness. It is amplified by Tr_1 and unbalances the bridge circuit. The voltage difference so produced between the emitter junctions of Tr_1, Tr_2, is continuously recorded on a suitable instrument.

To prevent polarization, polarity of the electrodes is reversed once per sec by the switch S_1, operated by a cam C, driven by the recorder mechanism. The electrode circuits are selected by a multipoint switch (not shown on the diagram) and it too is driven by the recorder.

To operate, switch S_2 is set to position 1, and the bridge circuit is balanced by potentiometer VR_1. The bridge has ranges from infinity to 100 k ohms, and infinity to 4 M ohms, and the output is 12 mV for full scale deflection of the recorder. The sensitivity required depends on the wetting properties of the leaf surface and the type of electrodes employed. The bridge circuit sensitivity depends on the supply voltage, which is set by the switch S_3, position 1 high, position 2 low; and finely adjusted by the potentiometer VR_2. The power unit, stabilized by the reference diodes R_1 and R_2, has an output of 16 V. The bridge is calibrated by switch S_2, positions 2 to 4, which provide a range of standard resistances. In position 5, the leaf electrodes are in circuit and the system records.

Figure 8a shows a record of the surface resistance of leaves at two heights in an irrigated and non-irrigated crop of field beans, taken during the night of 27 to 28 July 1966 when dew occurred. On the non-irrigated plot the crop height was 130 cm with almost full cover but sparse in places, and the soil moisture deficit was 54 mm: the irrigated crop had a full cover, was more luxuriant, being about 30 per cent denser, its height was 143 cm, and the soil moisture deficit was 26 mm.

The record from the mechanical dew-balance of dew on a single bean plant in the non-irrigated plot, during the same period, is in Fig. 8b. More dew was deposited in the irrigated plot largely because of the steeper vapour-pressure gradients within and above the crop, a usual feature of dense crops (Long 1958), and partly because the soil was wetter.

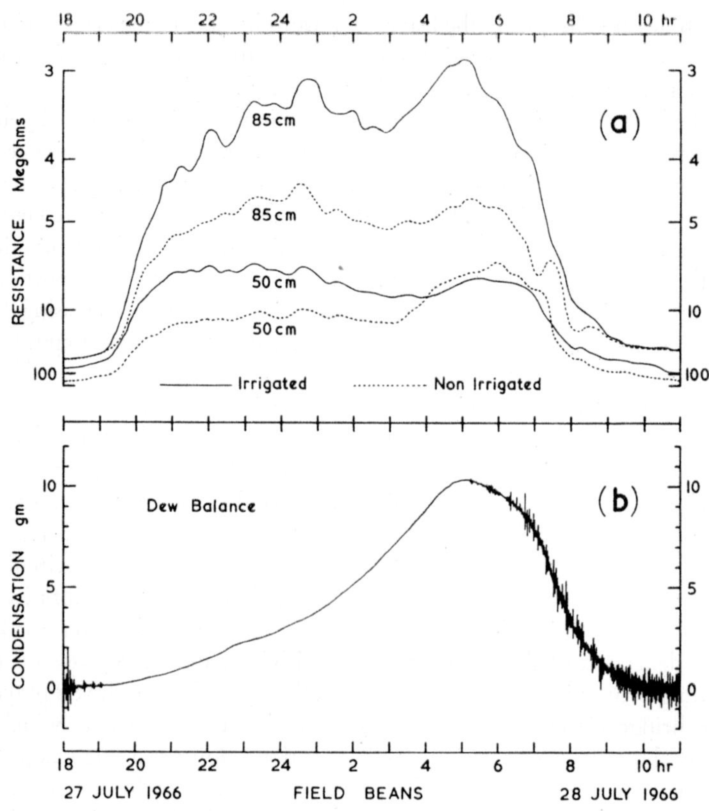

FIG. 8. Record of dew. (a) Surface-wetness of leaves. (b) Dew-balance record.

6 *Soil moisture*

Soil moisture is conveniently measured by the neutron scattering technique. The principle is simple. High energy neutrons from a suitable source are moderated to thermal neutrons in a soil medium, mainly by collisions with the hydrogen nucleus of water molecules. In field measurements, a probe containing a high energy neutron source and a thermal neutron detector are lowered down an access tube into the soil. The resulting thermal neutron flux is measured and, with suitable calibration, gives the volumetric moisture content of the soil. With a correctly designed probe in a homogenous soil the relationship is almost linear, but in the presence of a wetting front or moisture gradient, there is a tendency to underestimate the total moisture in a profile.

A portable neutron moisture meter suitable for field work (Long & French 1967) weighs less than 5 kg, including self-contained power supply. The unit is used to measure soil moisture profiles throughout the season in crops. The profiles indicate where roots are active. The net changes in soil moisture are calculated by integrating between sequential profiles and, after allowance for rainfall and percolation, the soil moisture deficit and total evaporation are estimated. These data are used to check evaporation rates calculated from measurements of the meteorological factors.

7 *Air-flow*

(i) *Sensitive cup-anemometers*

The air-flow above the crop canopy is best measured by small sensitive cup-anemometers of the type developed by Sheppard (1940). These instruments have nearly linear response at air-speeds down to 30 cm/sec, depart markedly from linear at slower air-speeds and stall around 15 cm/sec. For consistent results the anemometers require frequent calibration, particularly if they are used for profile or difference measurements, for which purpose they should, if possible, be carefully matched.

To measure air-flow profiles the anemometers are mounted on light telescopic masts, on side-arms extending about 30 cm. To prevent vibration and to keep the anemometer axis vertical, the mast is held rigid by light guy wires. The anemometers are mounted crosswind to the mast, where their performance is less influenced by the mast than when mounted downwind (Rider 1960).

Most recording versions of the sensitive anemometer are fitted with contacts (mechanical or photoelectric) giving one impulse per revolution of the cups. This rapid switching rate is ideal for short term recording, but for long term recording it is unnecessary and it shortens the life of the recording system. Very light mechanical contacts can be fitted onto the face of a standard counter-pattern sensitive anemometer (Casella T.16104) using a balanced 4-arm unit in place of the ordinary pointer. This is driven by the reduction gearing of the anemometer and operates the switch contacts to give one impulse for approximately 60 m run of the wind. The retarding action of the switch has no measurable effect upon the performance of the anemometer (for details, see Long 1957).

With the light contacts employed, a relay circuit is necessary between the anemometer and the recording system. If the contacts chatter, are dirty, or remain closed during a calm period, the relay unit must deliver only one pulse of fixed length to the recorder. A mains-powered circuit which satisfies these conditions is shown in Fig. 9a. It will operate with a contact resistance as great as one megohm, and, in the event of chattering contacts, will operate on the first closure and remain inoperative for a

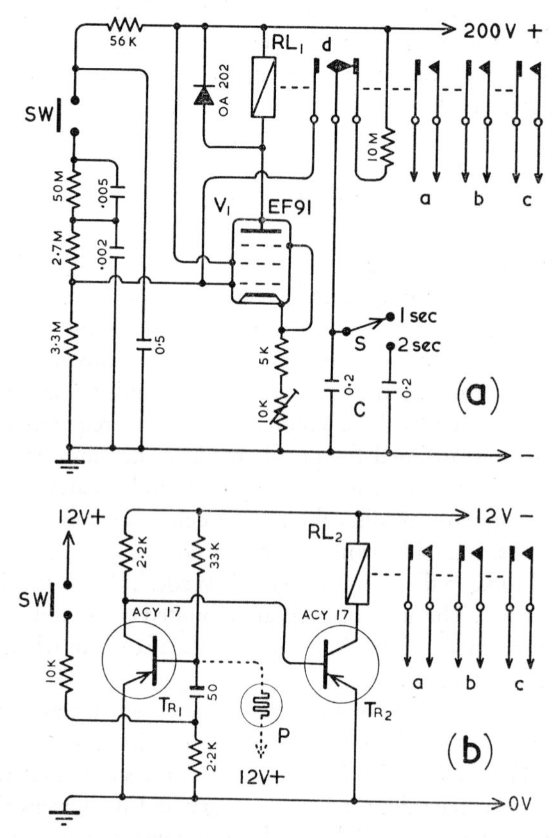

FIG. 9. Anemometer relay-circuits. (a) Mains version. (b) Battery version. All circuit values in ohms or μF unless specified. RL_1, 2000 ohms; RL_2, 160 ohms; P, photo conductive cell, ORP60.

period of 2 sec after the contact finally opens. (With a period of 2 sec between contacts a maximum airspeed of 30 m/sec can be handled.)

When the anemometer contact SW is closed the valve V_1 conducts momentarily, operating the relay RL_1, which closes the contact d. This connects a positive voltage to the grid of V_1, which conducts, and holds the relay on for a period determined by the discharge rate of the capacitor

C, selected by switch *S*. The 3 relay contacts *a*, *b* and *c* operate a counter for direct observation, a step-by-step chart recorder for pictorial presentation (Long 1951), and a 'print-out' data logging system, which gives the hourly totals of wind-run.

A battery-operated version of the relay unit (Fig. 9b) has similar properties to the mains driven unit, but contact resistance must not exceed 10 k ohm. If the control network between base and emitter of transistor Tr_1 is removed, the circuit can operate directly from a photoconductive cell, *P*. In this configuration, a counter replacing RL_2, can operate at speeds up to 50 counts/sec.

(ii) *Hot-bulb anemometer*

The performance of a cup anemometer deteriorates greatly near a surface and is quite unsuitable to measure air-flow within the crop canopy. To do this, a 4-point, constant current hot-bulb anemometer has been developed. It has a range of 0–200 cm/sec, and a response time of about 100 sec in an air-flow of 5 cm/sec, decreasing to about 30 sec at 200 cm/sec.

The instrument consists of two identical platinum thermometers, nominal resistance 100 ohms at 0°C, encapsulated in brass covers, (4 mm diam, 22 mm long). Around one thermometer is a heating coil (8·5 mm diam, 22 mm long, 15 turns) of 26 s.w.g. copper-nickel wire, resistance 1·4 ohm. The heater coil is supported in an anemometer shield so that the thermometer is central and not in contact with the coil. The shield consists of two thin brass discs (32 mm diam, 24 mm apart), separated by three copper wire supports (22 s.w.g.) One end of the coil is soldered to the upper disc and the other to a copper turret supported in the lower disc by an insulating bush (Fig. 10). The shield gives some protection from light rain, keeps the crops from coming into contact with the anemometer, and acts as a return path for current supply to the heater coil. The other thermometer is mounted in a standard radiation screen and compensates for changes in ambient temperature.

The two units are placed in opposition in a Wheatstone bridge circuit, indicated by the thick lines of Fig. 11. The anemometer unit is indicated by *e*, and the temperature compensation unit by *f*.

The heater coils of the four anemometer units are connected in series so that the power dissipated in each coil is the same. The current through the coils is kept constant at 1·8 amp by the mains-powered constant-current circuit *d*. The current is initially set by the potentiometer control R_1, and monitored by the ammeter *A*.

Power supply for the bridge is obtained from the voltage drop across the lead resistance Lr_1, (about 4 ohm for 200 m cable) and resistor R_7. The sensitivity of the circuit is preset by the variable resistor R_2.

To set the anemometer for use, the switch SW_1 is first switched to position 1, 'balance cold'. In this position the switch short-circuits the supply to the heater coils and brings R_4 into circuit, which compensates for the loss of bridge supply voltage (since in this position, Lr_1 and R_7

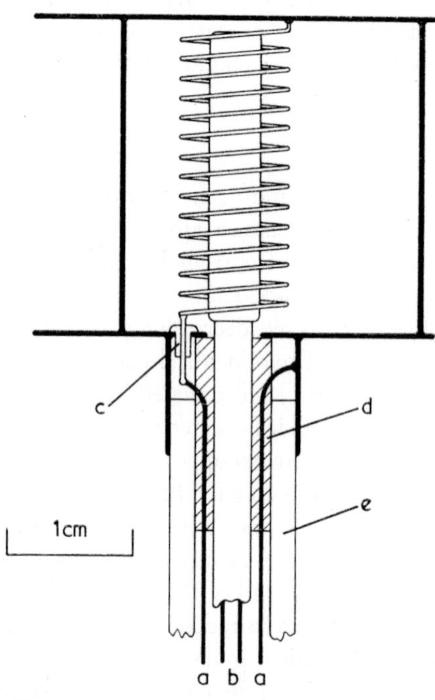

FIG. 10. Hot-bulb anemometer. (a) heater coil leads, (b) platinum thermometer leads, (c) insulating bush, (d) rubber-tubing, (e) Tufnol tubing.

are out of circuit). Resistor R_3 is also brought into circuit, replacing the load normally taken by the heater coils.

The four bridge circuits, selected by switch SW_2, are now individually balanced by adjusting potentiometers R_6 until there is zero deflection of the recording galvanometer G (100 μamp, 50 ohm). The anemometers are next shielded against air-flow, and SW_1 is switched to position 2, 'balance hot'. The heater current is now on; the anemometers heat up and the bridge becomes unbalanced. At equilibrium the out-of-balance current in each circuit is set for full scale deflection of the galvanometer

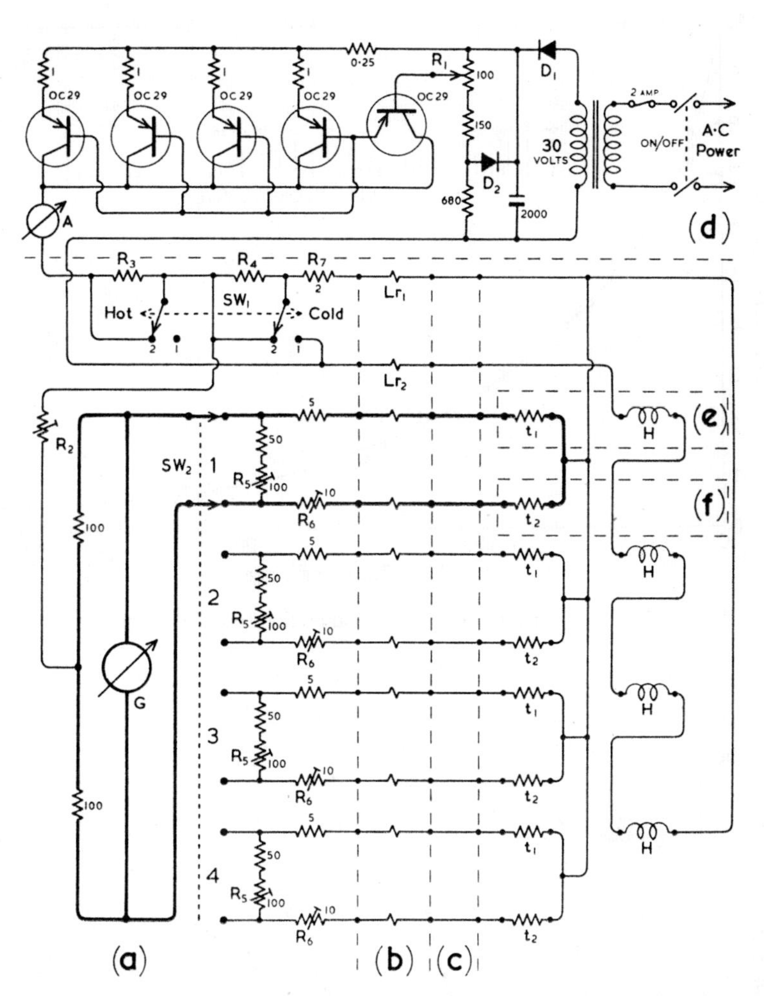

FIG. 11. Four-point, recording hot-bulb anemometer. (a) recorder unit, (b) field leads, (c) termination leads, (d) power unit, (e) hot-bulb anemometer, (f) ambient temperature compensation thermometer. All circuit values in ohms or μf. D_1, Rec 30; D_2, BZY96-C5V6.

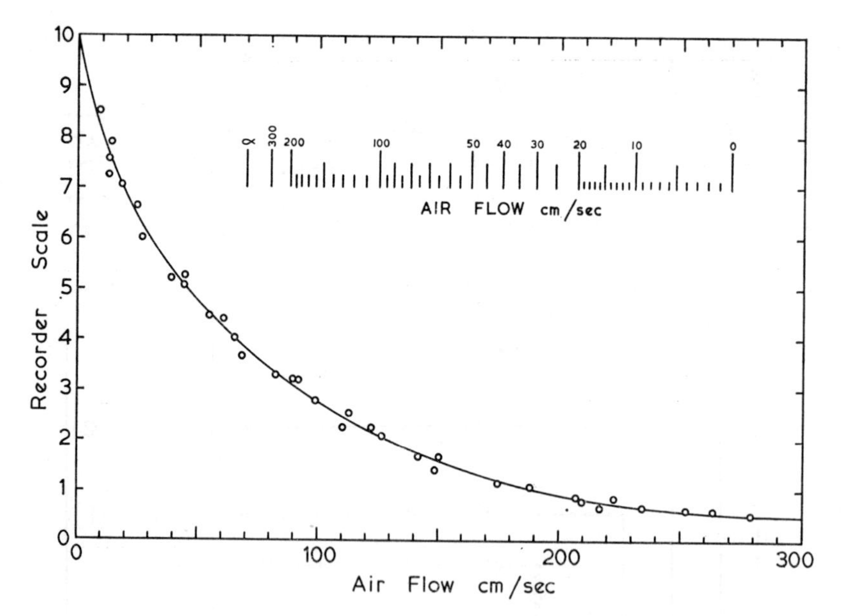

FIG. 12. Hot-bulb anemometer calibration, and scale form.

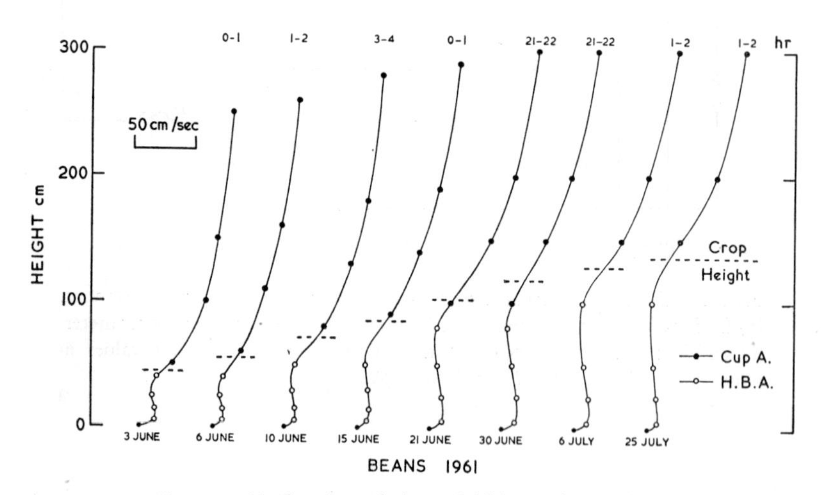

FIG. 13. Air-flow in and above field beans during dew.

G, by adjusting the shunting potentiometers R_5. The anemometers are now ready for use.

In an air current, the temperature change caused by heat loss from the hot-bulb unit tends to bring the bridge circuit back to balance, so decreasing the galvanometer deflection. This decrease, when suitably calibrated, is a measure of the air-flow. A field calibration against a standard cup-anemometer and a thermistor anemometer yielded the calibration curve of Fig. 12. Within the limits of calibration, about ± 5 per cent, the response of the anemometer is not affected by wind direction or by ambient temperature over the range 5–25°C. The form of the recorder scale is illustrated by the insert of Fig. 12. The response of the anemometer is non-linear, but can be made very nearly linear by reducing the diameter of the heater coil and by adjustment of component values in the bridge circuit. An earlier model had a linear range of 0–70 cm/sec, changing to non-linear above 70 cm/sec (Long 1957). The present instrument was designed to have an expanded scale at slow air-speeds to facilitate studies in dew formation.

Figure 13 gives examples of nocturnal air-flow profiles above and within a crop of field beans. All profiles are for periods of maximum dew deposition, throughout the growing season. The air-flow inversion within the crop coincides with the vapour pressure inversion, and is probably a result of the denser vegetation at this level. The S form of the profile persists throughout the day as the wind speed increases, and has also been recorded in barley and wheat (Long, Monteith, Penman & Szeicz 1964).

ACKNOWLEDGMENTS

I am grateful to Dr H.L.Penman and Mr B.K.French for advice and co-operation.

REFERENCES

Brooks F.A. (1959) *An introduction to Physical Microclimatology*. p. 93. Univ. Calif, Davis.

Brooks F.A. (Ed.) (1963) *Investigations of Energy and Mass Transfers near the Ground Including the Influences of the Soil-Plant-Atmosphere System*. p. 281–3. Univ. Calif, Davis.

Franssila M. (1936) Microklimatische Untersuchungen des Warmehaushalts. *Ilmatlet. Keskusl. Toim.* No. 20, 103 pp.

Hirst J.M. (1954) A method of recording the formation and persistence of water deposits on plant shoots. *Q. Jl R. met. Soc.* **80**, 227–31.

Hirst J.M., Long I.F. & Penman H.L. (1954) Micrometeorology in the potato crop. *Proc. Toronto Met. Conf.* 1953, 233–7.

Long I.F. (1951) Automatic run-of-wind recorder. *Proc. Instn elect. Engrs* **98**, Pt II, 458.

Long I.F. (1955) Dew and guttation. *Weather, Lond.* **10**, 128.

Long I.F. (1957) Instruments for micro-meteorology. *Q. Jl R. met. Soc.* **83**, 202–14.

Long I.F. (1958) Some observations on dew. *Met. Mag., Lond.* **87**, 161–8.

Long I.F. & French B.K. (1967) Measurement of soil moisture in the field by neutron moderation. *J. Soil Sci.* **18**, 149–66.

Long I.F., Monteith J.L., Penman H.L. & Szeicz G. (1964) The plant and its environment. *Met. Rdsch.* **17**, 97–101.

McIlroy I.C. (1955) The atmospheric fine structure recorder. *Tech. Pap. Div. met. Phys. C.S.I.R.O. Aust.* No. 3, 1–19.

Middleton W.E. & Spilhaus A.F. (1953) *Meteorological Instruments*, 3rd Edition, p. 85–8. Univ. Toronto Press.

Monteith J.L. & Szeicz G. (1962) Simple devices for radiation measurement and integration. *Arch. Met. Geophys. Bioklim.* **B.11**, 491–500.

Penman H.L. (1963) Vegetation and hydrology. *Tech. Commun. Commonw. Bur. Soil Sci.* No. 53, 124 pp.

Penman H.L. & Long I.F. (1949) A portable thermistor bridge for micrometeorology among growing crops. *J. scient. Instrum.* **26**, 77–80.

Penman H.L. & Long I.F. (1960) Weather in wheat: an essay in micrometeorology. *Q. Jl R. met. Soc.* **86**, 16–50.

Platt R.B. & Griffiths J.F. (1964) *Environmental Measurement and Interpretation*, p. 93–7. Chapman and Hall, London.

Rider N.E. (1960) On the performance of sensitive cup anemometers. *Met. Mag., Lond.* **89**, 209–15.

Robinson N. (1966) *Solar Radiation*, p. 222–316. Elsevier, London.

Sheppard P.A. (1940) An improved design of cup anemometer. *J. scient. Instrum.* **17**, 218–21.

Szeicz G., Monteith J.L. & Dos Santos J.M. (1964) Tube solarimeter to measure radiation among plants. *J. appl. Ecol.* **1**, 169–74.

Tanner C.B. (1963) Basic instrumentation and measurements for plant environment and micrometeorology. *Soils Bull.* No. 6. Univ. Wisconsin, Madison.

Wexler A. (Ed.) (1965) *Humidity and Moisture*, vol. 2. (ed. E.J.Amdur), Reinhold, New York. 634 pp.

THE MEASUREMENT OF TEMPERATURE, HUMIDITY AND CARBON DIOXIDE/OXYGEN LEVEL AMONGST STORED FOOD PRODUCTS

F.L. WATERHOUSE AND T.G. AMOS*

Department of Natural History, University of Dundee (formerly: Queen's College, University of St. Andrews)

SUMMARY The measurement of temperature, humidity and carbon dioxide/oxygen levels amongst stored produce is discussed. Temperature may be determined using a specially designed thermocouple spear which permits measurement at almost any desired location in the stored produce. The use of various psychrometric expressions for humidity is discussed. A description and appraisal of some humidity measuring techniques is given with special attention centred on conductivity sensors. An indirect approach to measuring humidity is to determine the moisture content of the produce and then relate this to humidity by means of humidity equilibria curves. The measurement of carbon dioxide/oxygen concentration as an indicator of the degree of insect infestation can not be accomplished satisfactorily in bagged produce and only in bulk if samples of grain are taken for gasometric analysis.

INTRODUCTION

Temperature, humidity and to a lesser extent the level of carbon dioxide and/or oxygen are important factors of the microclimate which influences the behaviour and biology of insect pests in stored food products. The need to measure these factors prompted the development of various techniques some of which are outlined below. In Britain, the Pest Infestation Laboratory, Slough, and the Infestation Control Laboratory, Tolworth, have carried out work on these problems. These techniques are

* On transfer from Tropical Stored Products Centre, Ministry of Overseas Development, Slough, Bucks.

necessarily limited by the conditions prevailing but opportunity is taken to discuss the underlying principles involved. It is thus hoped to help others to adapt these techniques to meet their own specific requirements. In general, measurements made only in situations where insolation is not an immediate environmental component will be considered. The shielding of instruments from solar radiation with its consequent hazard of providing local climates around the sensors will not be discussed.

TEMPERATURE

The temperature at the surface and in the peripheral regions (say, outer 30 cm) of bulk and bagged produce may be obtained using a variety of instruments ranging from mercury-in-glass, mercury-in-steel (Bourdon type) and bimetal strip thermometers to electrical resistance thermometers, thermistors and thermocouples. The choice of a particular technique will

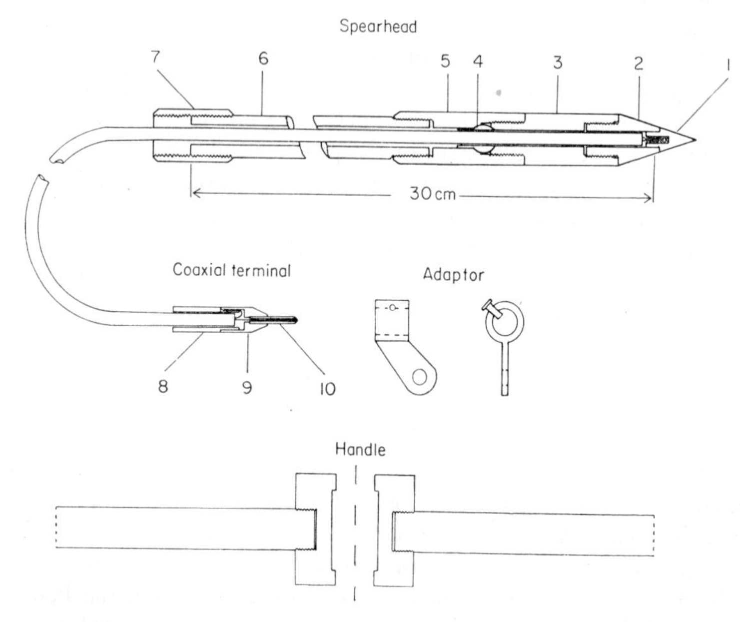

FIG. 1. The thermocouple spear; 1, brass tip incorporating the thermojunction; 2, thermal insulator; 3, first part of body; 4, internal split collet which holds thermocouple wire; 5, second part of body; 6, remainder of spearhead body; 7, chamfered steel coupling; 8, brass sleeve connection for copper wire; 9, insulator; 10, connection for eureka wire. (From Haswell & Oxley 1957.)

depend on a variety of considerations most of which have been discussed by Siddorn (1961).

The measurement of temperature within bulk or stacks of bagged produce not specially prepared for such measurement presents a problem which has been largely overcome by employing a thermocouple spear* (Haswell & Oxley 1957). The spear (Fig. 1) consists of: a spearhead, a shaft, a co-axial terminal and a handle. The spearhead has a small brass tip 12·7 mm long incorporating the thermojunction which is thermally insulated from the first part of the spearhead of 17·3 mm diam, the remainder being nominal $\frac{1}{4}$ in. (6·4 mm) gas piping. The length of the spearhead is such that the thermojunction included in the brass tip is 30 cm from the end of the gas piping. The shaft is composed of 30 cm lengths of $\frac{1}{4}$ in. gas piping joined by chamfered steel couplings. The co-axial terminal is designed so that it can be easily threaded through the lengths of piping before these are connected to the spearhead or shaft. A special adaptor is necessary to connect this terminal to the normal screw cap terminals on the commercial Doran mini potentiometer.† The handle consists of two brass blocks hinged together so as to form, when closed, a composite block. This block is bored along the mid-line of the two contiguous faces of its two halves, with a hole to take the gas piping in the region of the coupling joining two lengths of shaft. By means of two steel side arms, one from each outer face of the composite block, the handle is latched around a coupling and pressure applied to the side arms.

The spear may be inserted up to a depth of about 6 m into bagged produce and perhaps a little further into bulk. By noting the number of 30 cm lengths of shafts used, the position of the speartip within the produce can be determined. In bulk produce, individual speartips may be detached and left *in situ* for convenience of later readings. This instrument has been used successfully here and abroad.

HUMIDITY

A major problem in biological work is the measurement of humidity coupled with the choice of a suitable psychrometric unit for the interpretation of results. Opinions differ as to which of the usual methods of expressing atmospheric humidity, i.e. relative humidity (a ratio of vapour pressures) and saturation deficit (a difference between vapour pressures)

* Made by B. and M. Electronics Ltd., Crawley, Sussex.

† Made by Doran Instruments Ltd., Stroud, Gloucestershire.

is more appropriate. Although relative humidity and saturation deficit are defined terms:

$$\text{saturation deficit (S.D.)} = e_a - e$$
$$\text{relative humidity (R.H.)} = e/e_a$$

where e= actual vapour pressure at a given temperature and
e_a= saturation vapour pressure at the *same* temperature

there seems to be some confusion in not only their relationship to temperature, but also in their applicability to biological phenomena such as water loss.

It is proposed therefore to consider briefly this problem before describing some of the techniques used for measuring humidity in stored products work. According to the literature one examines and ones own attitude of mind, the relationships between R.H. and S.D. can be obvious or obscure. So also can the part the atmospheric moisture conditions so described play in biological phenomena such as water loss. In most cases lack of clarity stems from confusing mathematical descriptions of atmospheric moisture conditions with the importance or otherwise of those conditions in relation to some event, i.e. on the one hand descriptions and on the other functional potentialities of conditions described.

Both descriptions, R.H. and S.D., concern the moisture content of the atmosphere but further information is required before they can be related; and this is temperature. Thus, to the first decimal place, at 10°C, the term 19·6 per cent R.H. describes the same moisture condition of the atmosphere as does 7·4 mm of Hg S.D. At different temperatures however the similarity of moisture conditions in certain aspects is more easily recognized by one formula than the other. Thus at 20°C with an R.H. of 57·7 per cent the S.D. is still only 7·4 mm of Hg. Because air at 20°C can hold more moisture than at 10°C the R.H. ratio has altered although the pressure of water vapour still required to saturate the atmosphere (S.D.) is the same. This latter fact does not emerge from the R.H. description of the moisture conditions unless the appropriate calculation is made. Atmospheric pressure can be important but usually does not vary to the degree that it must be considered when calculating R.H. or S.D. in any given area, however see Marvin (1941).

Turning now to the functional potentialities of these atmospheric moisture conditions, the choice of descriptive terms depends upon the potentials one wishes to examine. In the case of water loss the evaporating power of the air is generally considered to be important. However it has been said that it is the vapour pressure gradient between the evaporating surface and the air which determines this physical process which is not therefore directly related to either the R.H. or S.D. of the air (Dethier 1963). Many people might consider this to be begging the question. However there is certainly some argument as to the extent to which either of the two methods of description assesses the evaporating power of the air. Wigglesworth (1965) quotes Dalton's law:—

$$V = a.E(100 - H) + c \qquad \text{(1)}$$

where V is rate of evaporation,
E is saturated vapour pressure at the temperature in question,
H is relative humidity,
a and c are constants,

relating to the rate of evaporation from a water surface and states that it is applicable to the rate of water loss in insects. He then goes on to mention many exceptions some but not all of which are shown to depend upon biological phenomena. Ramsay (1935) draws particular attention to the importance of air movement, the increase in the rate of diffusion at high temperatures, and the relative temperatures of the evaporating body and the atmosphere (presumably applicable to insolated bodies such as animals and plants exposed to sunshine). He points out that whilst the law may be true under certain conditions it is an incomplete statement of fact and may be misleading. Edney (1957) comments similarly. The law as quoted is therefore at best only a major factor contributing to water loss in the particular biological system under consideration but it does involve the descriptive terms since:

$$\text{S.D.} = E(100 - H)$$

where E is saturated vapour pressure and H is relative humidity ratio expressed as percentage. This is perhaps better expressed:

$$\text{S.D.} = \text{E}(1 - h)$$

where h is relative humidity expressed as a simple ratio (Penman 1955).

In the law (equation 1) relating to the special case of still air and free water surfaces, the term $E(100 - H)$, which is the S.D. calculated using saturated vapour pressure and R.H., is the only variable. It is understandable therefore that evaporation has been considered to be directly proportional to S.D. Hence the statements by Buxton (1932), Mellanby (1932) and others to this effect although other contributive circumstances were later recognized. Beament (1958) in spite of his recognition of its limitations opts for S.D. as the most useful biological expression of humidity. Indeed whatever the peculiarities of the plant or animal involved, S.D. must play a major role as far as the atmosphere is concerned although it may not achieve the expected effect. It is also worth noting and contrary to the belief of many that the functional potentials of a moisture condition indicated by S.D. are more fully described if relevant temperatures are quoted. Here the remark above concerning temperature and rates of gaseous diffusion is relevant (Ramsay, 1935). In temporarily accepting this situation it is therefore all the more important to treat correlations of S.D. and R.H. with vital phenomena, cautiously. Indeed in some cases other factors are suggested to be involved to the extent of completely offsetting the usually important role of S.D. and R.H. in determining evaporation conditions attractive to certain insects (Wellington 1949). Similarly insect physiologists have claimed that the responses by sense organs are correlated to R.H. rather than S.D. and vice versa (Pielou 1940; Lees 1943, among others).

The techniques available for measuring humidity are numerous and varied (Wexler & Brombacher 1951; Wexler 1965; Gregory & Rourke 1957; Hickman 1958; Penman 1955) but only small instruments are suitable for use in confined spaces. An excellent account of the techniques applicable to ecological work is given by Macfadyen (1963) and here it is therefore only necessary to mention our special experience with some of these techniques and to describe an indirect method depending upon the relationship between the moisture content of the medium and R.H. of the atmosphere with which it is in equilibrium. Many of the adverse qualities of the small instruments now to be discussed do not apply to larger versions operating in unenclosed spaces.

The Assmann aspiration psychrometer and sling psychrometers may be used extensively to record relative humidity in the general atmosphere of warehouses. In small spaces however, even the thermocouple and thermistor models have their drawbacks, although the reduction in size of the temperature sensors permits a reduced air flow and an absolute reduction of the water vapour introduced into the atmosphere under investigation.

Such instruments may be satisfactory in some outdoor locations, for example amongst grass where an open system of spaces ensures that a suction speed of 5 cm/sec merely samples air from similar thermal strata of comparable humidity (Waterhouse 1955). Indeed an Assmann sucking at 4 cm/sec would have given almost comparable results. When closed spaces of small dimensions are under inspection however the question of where the air comes from and how much the atmosphere of the space is altered by the introduced water vapour becomes paramount. Even when the forced draught is excluded and the instrument is calibrated under these conditions (for example, Powell 1936), the introduced vapour may present hazards in very small spaces and if the spaces are open and an unmeasured natural draught occurs the experimenter is at a loss to know which reference tables to use to convert temperature depressions to relative humidity. Such instruments may serve however if very limited accuracy is required and on special occasions, for example the psychrometer described by Monteith & Owen (1958) may serve where the range 95–100 per cent R.H. only is of interest. In all these cases, however, the provision of a water reservoir and a wetting device increases sensor bulk or if held elsewhere and connected by conduit, introduces temperature differentials.

A technique which owes much to the work of Dunmore (1938), Gregory (1946) and Koie (1948) requires measurement of the electrical conductivity of a substance which absorbs or surrenders moisture according to the humidity of the surrounding air. Typically the sensor consists of platinum or rhodium electrodes round which is wound a continuous spiral thread of glass fibre impregnated with a 1–2 per cent aqueous solution of calcium (or lithium) chloride. The resistivity of the glass fibre thread will thus vary with relative humidity. There are a number of difficulties with this method in that the resistance is also dependent on temperature, that there is a tendency for the hygroscopic salt to be 'leached' out at high humidities, that there is a hysteresis effect and the overall characteristics of the sensor change with age, thus necessitating frequent recalibration. Moreover it cannot be used in atmospheres containing vapours of ammonia, formic acid etc., as these substances react with the impregnated salt and thus destroy the sensor's calibration (Peacock, Waterhouse & Baxter 1955). This may also be true of other techniques employing a chemical reagent, for example cobalt thiocyanate papers (Solomon 1945, 1957). The value of this conductivity sensor is thus limited though it has been used successfully by for example, Edney (1952) and Amos & Waterhouse (1967). Small conductivity sensors, or 'humistors', produced commercially usually consist of two interdigitated electrodes printed on the surface of a 'plastic' base measuring approximately 22 mm wide × 42 mm long × 1 mm thick. The printed electrodes are chemically protected against abrasion and contamination by various forms of foreign matter (for example, dust particles) and enclosed in a perforated case which affords protection against rough handling etc. It is interesting to note that Eighme (1965, personal communication) found that some sensors of a similar type, when placed amongst stored produce, were damaged as a result of insects eating away parts of the printed electrodes.

In all instruments of this type the sensor is connected to a battery operated meter and calibrated directly in relative humidity. The problem of dealing with the enormous variation in sensor resistance over a wide humidity range usually involves the use of more than one scale and a selector switch.

Measurement of low and high humidities may reveal considerable discrepancies between the meter readings and the true humidity (as determined by an Assmann psychrometer), particularly at extreme conditions. For example, between 40–50 per cent R.H. one instrument indicated values 10–20 per cent R.H. too low and below 40 per cent R.H. the readings were completely unreliable. Further, the instrument was unreliable at humidities about 90 per cent R.H. and above. Eighme (1965 personal communication) reported variations of up to 5 per cent R.H. during a period of 6 months due to damage by saturated conditions to a similar type of sensor. This sensor was also found to have a maximum hysteresis effect of $\pm 2{\cdot}5$ per cent R.H. at 50 per cent R.H. (Walker & Turner 1964). These variations were all accentuated by low temperatures (ranging from 13–26°c).

During the course of these observations it became evident that there might not be a continuous reading between the two meter scales. In one case a reading of 59 per cent R.H. on the lower scale was equivalent to 65 per cent R.H. on the upper scale. It was subsequently found that readings in the range 56.5–58 per cent R.H. were all too low but could be corrected, whereas for readings greater than 58 per cent R.H. it was necessary to revert to the high scale. Thus 58 per cent R.H. on the lower scale was equivalent to 60 per cent R.H. on the high scale which reflected the true relative humidity. It was thus impossible to obtain meter readings in the 58–60 per cent R.H. range.

Sensors were found to respond rapidly (within 10–20 sec) to changes in humidity but would only approximate to their equilibrium levels within a matter of minutes. Complete equilibrium could only be expected after a considerably longer period had elapsed. At low humidities the final reading could be taken within 20–30 min whereas very much longer periods were required in high humidities. Consequently it was difficult to assess the relative humidity in fluctuating conditions.

The temperature correction required with this type of instrument varies with the design but is usually considerable and may not be constant. In our own experience with various types of sensor design, matching over a wide humidity range was extremely difficult to achieve if one required better than ±4 per cent comparability. However whilst some individual sensors have been found to show error of as much as ±8–10 per cent R.H. over a wide range of humidity, this does not necessarily prohibit their use in more limited circumstances. It is possible to achieve great accuracy if only *relative* changes in humidity are to be noted. Thus one can employ numerous elements each working over a very limited range (Amos 1965, Amos & Waterhouse 1967). Under these circumstances frequent recalibration is an accepted chore.

With respect to other techniques it is worth appraising those instruments where a sensitive hygroscopic element of hair or paper, which changes shape and in doing so moves a lever across a scale, is located within a perforated flat box. The commercial Edney, direct reading, paper and hair hygrometers fall into this category but devices operating on this principle have been made privately (Howell & Craig 1939). First the whole instrument must be located in the microclimate and the possibility exists, under changing conditions, that the sample of air within the box may not follow very precisely that outside. The casing itself may introduce a temperature lag. Then the scale at the high humidity end is cramped and this becomes more serious as the instrument becomes smaller. Furthermore hysteresis is prone to occur with frequent change between extreme conditions and results in the instrument having to be recalibrated, as also does rough usage. In limited circumstances however, for example where small changes occur within a reduced range, useful results can be obtained providing the instruments are stored in humidity conditions within their operational zone. In any case a frequent calibration check at a good number of points within the required range is advisable with these instruments as with the standard recording hair hygrometers in general laboratory use if absolute and not relative readings are desired. In such cases the customary one point check at current ambient conditions against an Assmann or a Whirling hygrometer must occasionally be substituted by a more thorough procedure.

Calibration is a subject in itself. Few really good test chambers exist or are available to researchers. For this reason the Tropical Stored Products Centre, Slough, Bucks. is constructing a test chamber as a prerequisite to a research programme aimed at providing better humidity sensors. Simple chambers located in thermostatted tanks and containing air pulled from adjacent wet and dry sources can be set up in any laboratory and monitored by a wet and dry thermocouple procedure. On the other hand graded or saturated solutions (Solomon 1951) may be considered convenient. For accurate work constant temperature conditions and a stirring device for the air above the liquid are necessary but can be abandoned during rough checks in the field.

An indirect method for determining the humidity amongst stored produce is to determine the moisture content of the commodity in question since all commodities have equilibria characteristics between the moisture contained and the water vapour in the associated air. For each type and variety of food grain there is a characteristic equilibrium curve (Davey & Elcoate 1965, 1966), or humidity balance curve, which is

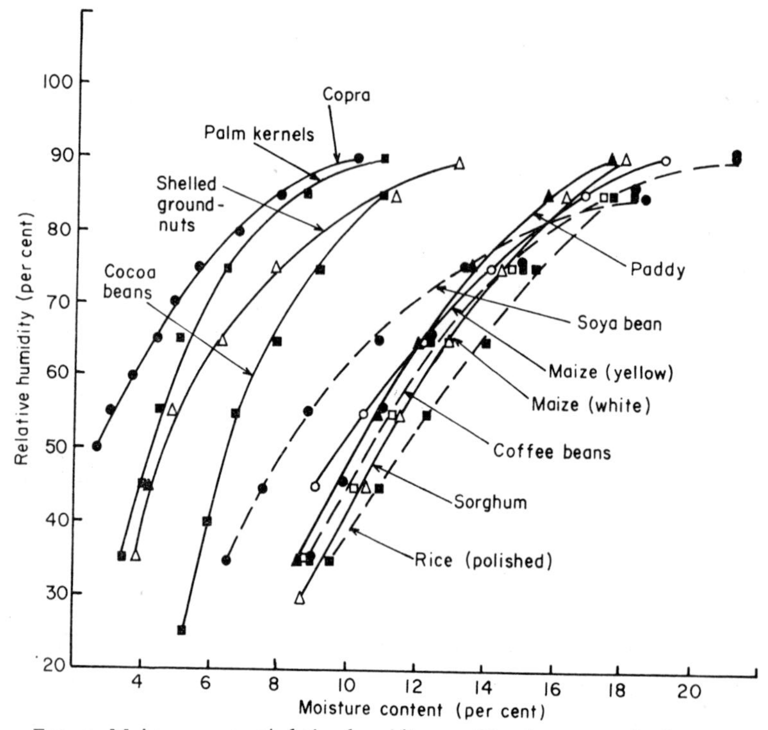

FIG. 2. Moisture content/relative humidity equilibration curves (each curve has been compiled from results of several workers to present a guide to the moisture content/relative humidity equilibration values, which vary between varieties of the same species of food product). (from Hall D.W. 1963. Informal Working Bulletin No. 24, Agricultural Engineering Branch, Food and Agriculture Organisation of the United Nations.)

obtained by plotting a graph of the moisture content against the relative humidity of the air (Fig. 2). By determining the moisture content of the commodity it is therefore possible to assess the relative humidity of air with which it is in association.

Moisture content measuring devices are many and varied. Laboratory methods include infra-red techniques and the standard reference oven drying and oil distillation procedures. Field apparatus ranges from hair hygrometers to chemical (acetylene) and battery operated (electrical resistance and capacitance) meters. The temperature of the commodity must be ascertained during the course of the moisture content determination when an electrical meter is used so that the moisture content value may be corrected accordingly.

Moisture content determinations at the periphery of stacks of bagged produce may be carried out using an electrical probe type meter which can be thrust into the outer regions (say 30 cm). Alternatively, a small sample may be taken by means of a bag sampling spear and either analysed on the spot using a portable field moisture meter or sealed in a moisture proof vessel and taken back to the laboratory for analysis. Determinations within stacks of baggged produce are not possible. Where produce is stored in bulk, it is possible to obtain samples by means of bulk or cargo sampling spears. Such spears have a number of disadvantages most of which are overcome by the relatively new suction type spear (Burges 1960). This spear can be inserted into bulk in any direction, even upwards. When sampling downwards, samples can be obtained from as deep as 9 m and from directly against the walls of a store as close as 1 cm to the floor. Again, moisture content determinations can be carried out either on the spot or at the laboratory. The sampling of a commodity raises various problems (Warner 1955) which will not be dealt with here except to recall that removal of large samples from bulk will cause subsidence and certainly prevent sequential sampling over a series of visits at intervals of time (Howe 1964). This indirect approach to determining the humidity amongst stored produce is a valuable technique providing the inherent difficulties are fully appreciated.

CARBON DIOXIDE/OXYGEN

Knowledge of the carbon dioxide/oxygen concentration in stored produce may give an indication of general keeping qualities. In a hermetically sealed storage container, the insect infestation in the grain will gradually die out as it depletes the oxygen with a corresponding increase in carbon dioxide. In non-airtight storage containers, however, it becomes necessary to assess from time to time the degree of infestation present. The detection of insects in such circumstances may be achieved indirectly by measruing the carbon dioxide level to which the degree of infestation is related. It is only in exceptional circumstances that the respiration of micro-organisms in the produce itself will be of the same order as even a light insect infestation. The measurement of carbon dioxide/oxygen within stored produce presents a problem which has not been satisfactorily overcome, primarily due to the lack of a simple and inexpensive sensor which is capable of recording at a remote point. For this reason the general practice is to remove grain samples which are then sealed in airtight containers and taken to the laboratory where they are incubated at a standard temperature for a standard time. A gasometric method accurate to $\pm 0{\cdot}2$ per cent carbon dioxide, designed by Oxley & Howe (1952) especially for the analysis of grain samples for carbon dioxide, is used. Other apparatus which gives carbon dioxide concentrations on a volumetric basis with sufficient accuracy may be used, e.g. a standard Haldane apparatus. For rough estimates of the carbon dioxide/oxygen concentrations, Fyrite gas analysers* may be adequate. This method is one of many which may be used for detecting insects in stored produce (Ashman 1966).

ACKNOWLEDGMENTS

The authors are indebted to Dr D.W.Hall, Director of the Tropical Stored Products Centre, Ministry of Overseas Development, Slough, and Mr G.V.B.Herford, C.B.E., Director of the Pest Infestation Laboratory, Agricultural Research Council, Slough, Bucks. for permission to use Figures 1 and 2; also to Mr J.H.Hammond of the Pest Infestation Laboratory for providing photographs of the figures. They also wish to express gratitude to Drs W.M.Graham (Tropical Stored Products Centre) and A.K.Onyearu (Nigerian Stored Products Research Unit) for many useful discussions and to Messrs P.E.Wheatley and P.J.Mackay (Tropical Stored Products Centre) for reading the manuscript.

* From Shandon Scientific Co. Ltd., Willesden, London, N.W.10.

REFERENCES

AMOS T.G. (1965) Studies on the orientation behaviour of *Carpophilus dimidiatus* (F.), *C. hemipterus* (L.) and other stored products beetles, in humidity gradients. Ph.D. Thesis. University of St. Andrews.

AMOS T.G. & WATERHOUSE F.L. (1967) Phasic behaviour shown by two *Carpophilus* species (Coleoptera, Nitidulidae) in various humidity gradients and its ecological significance. *Oikos* **18**, 345–50.

ASHMAN F. (1966) Inspection methods for detecting insects in stored produce. *Tropical Stored Products Information* **12**, 481–94.

BEAMENT J.W.L. (1958) Measurement and control of humidity. In: *Electronic Apparatus for Biological Research* (Ed. P.E.K.Donaldson). p. 413–18. Butterworth.

BURGES H.D. (1960) A spear for sampling bulk grain by suction. *Bull. ent. Res.* **51**, 1–5.

BUXTON P.A. (1932) Terrestrial insects and the humidity of the environment. *Biol. Rev.* **7**, 275–320.

DAVEY P.M. & ELCOATE S. (1966) Moisture content/relative humidity equilibria of tropical stored produce (Part I Cereals). *Tropical Stored Products Information* **11**, 439–67.

DAVEY P.M. & ELCOATE S. (1965) Moisture content/relative humidity equilibria of tropical stored produce (Part 2 Oilseeds). *Tropical Stored Products Information* **12**, 495–512.

DETHIER V.G. (1963) *The Physiology of Insect Senses*. Methuen.

DUNMORE F.W. (1938) An electrical hygrometer and its application to radio meterography. *J. Res. Natl. Bur. Stan.* **20**, 723–44.

EDNEY E.B. (1952) An electrical hygrometer suitable for microclimatic measurements. *Trans. IX Int. Congr. Ent. Birmingham* **1**, 525–30.

EDNEY E.B. (1957) *The Water Relations of Terrestrial Arthropods*. Cambridge.

GREGORY H.S. (1946) The measurement of humidity. *Instr. Pract.* **1**, 447.

GREGORY H.S. & ROURKE E. (1957) *Hygrometry*. Crosby Lockwood.

HASWELL G.A. & OXLEY T.A. (1957) A thermocouple spear for temperature measurement in stacks of bagged produce. *Trop. Agric.* **34**, 55–8.

HICKMAN M.J. (1958) Measurement of humidity. *National Physical Laboratory Notes on Applied Science* No. 4 H.M.S.O. London.

HOWE R.W. (1964) Investigation of the efficiency of spear sampling of bulk grain. *Proc. XII Int. Congr. Ent. London* 629.

HOWELL D.E. & CRAIG R. (1939) A small hygrometer. *Science* **89**, 544.

KØIE M. (1948) A portable alternating current bridge and its use for microclimate temperature and humidity measurements. *J. Ecol.* **36**, 269–82.

LEES A.D. (1943) On the behaviour of wire worms of the genus *Agriotes* Esch (Coleoptera, Elateridae) 1. Reactions to humidity. *J. exp. Biol.* **20**, 43–53.

MACFADYEN A. (1963) *Animal Ecology, Aims and Methods*. 2nd edition. Pitman.

MARVIN C.F. (1941) Psychrometric tables for obtaining the vapour pressure, relative humidity, and temperature of the dew point. *U.S. Dept. Comm., W.B. No.* 235.

MELLANBY K. (1932) The effect of atmospheric humidity on the metabolism of the fasting mealworm (*Tenebrio molitor* L. Coleoptera). *Proc. roy. Soc. Lond.* B. **111**, 376–90.

MONTEITH J.L. & OWEN P.C. (1958) A thermocouple method for measuring relative humidity in the range 95–100%. *J. Sci. Instr.* **35**, 443–46.

OXLEY T.A. & HOWE R.W. (1952) *Detection of Insects by their Carbon Dioxide Production.* H.M.S.O. London.

PEACOCK A.D., WATERHOUSE F.L. & BAXTER A.T. (1955) Studies in Pharaoh's ant, *Monomorium pharaonis* (L.) 10. Viability in regard to temperature and humidity. *Ent. mon. Mag.* **91**, 37–42.

PENMAN H.L. (1955) *Humidity.* Chapman & Hall.

PIELOU D.P. (1940) The humidity behaviour of the mealworm beetle, *Tenebrio molitor* L. II. The humidity receptors. *J. exp. Biol.* **17**, 295–306.

POWELL R.W. (1936) The use of thermocouples for psychrometric purposes. *Proc. Phys. Soc.* **48**, 406–14.

RAMSAY J.A. (1935) Methods of measuring the evaporation of water from animals. *J. exp. Biol.* **12**, 355–72.

SIDDORN J. (1961) Temperature measurement in stored products work. *Tropical Stored Products Information* **3**, 68–76.

SOLOMON M.E. (1945) The use of cobalt salts as indicators of humidity and moisture. *Ann. appl. Biol.* **32**, 75–85.

SOLOMON M.E. (1951) Control of humidity with potassium hydroxide, sulphuric acid and other solutions. *Bull. ent. Res.* **42**, 543–54.

SOLOMON M.E. (1957) Estimation of humidity with cobalt thiocyanate papers and permanent colour standards. *Bull. ent. Res.* **48**, 489–506.

WALKER P.T. & TURNER M.L. (1964). A small temperature and humidity meter based on thermistors and humistors. *Tropical Pesticides Research Unit, Porton Report No.* 271.

WARNER M.G.R. (1955) The sampling of grain. *The Agricultural Merchant*, May and June.

WATERHOUSE F.L. (1955) Microclimatological profiles in grass cover in relation to biological problems. *Quart. J. R. Met. Soc.* **81**, 63–71.

WELLINGTON W.H. (1949) The effect of temperature and moisture upon the behaviour of the spruce budworm, *Choristoneura fumiferana* Clemens (Lepidoptera, Tortricidae). 1. The relative importance of graded temperatures and rates of evaporation in producing aggregations of larvae. *Sci. Agric.* **29**, 201–15.

WEXLER A. (1965) *Humidity and Moisture* 1. *Principles and Methods of Measuring Humidity in Grass.* (Ed. A.Wexler.) Reinhold.

WEXLER A. & BROMBACHER W.G. (1951) Methods of measuring humidity and testing hygrometers. *U.S. National Bureau Standards, Circular No.* 512.

WIGGLESWORTH V.B. (1965) *The Principles of Insect Physiology.* 6th edition. Methuen.

THE MEASUREMENT OF APPARENT SURFACE TEMPERATURE

D. B. IDLE

Department of Botany, University of Birmingham

SUMMARY The fluxes of energy and matter to and from the leaf can be quantified by a variety of means, but these usually involve a knowledge of the temperature of the leaf and the air. The use of thermocouples and thermistors for measuring these temperatures is well known to be subject to error for a variety of reasons. The radiometric determination of source temperature is now a practical additional technique for the biologist. The errors involved in this and the traditional methods are discussed in relation to leaf structure, the purpose for which the temperature determination is made, and the 'worst case' environmental conditions. A simple radiometer for field use, sensitive to the 5–9 μ waveband, is described.

INTRODUCTION

The surface of a plant or animal is a region in which exchange phenomena involving heat and matter occur across a system that is not only physically heterogenous, but also includes a change of state in all but aquatic organisms. Anatomically there is usually no doubt where the surface lies, and in plants it might be supposed that the thin outer cuticular layer would help define the surface. But physiological studies involving exchange phenomena are concerned with the entire surface system of gradients and discontinuities, with the following consequences. Firstly, it becomes increasingly difficult to assign properties such as temperature and water content to the anatomical surface. Secondly, the physical size of the system is so small, and our measuring instruments need so much energy to work them compared to that available in the space immediately around them, that we have severe limitations on what it is possible to measure, as well as the usual errors due to interference with the system under investigation. Moreover, the gross anatomy of the leaf, which is of prime interest because it is the organ of exchange and synthesis, is such that while the major part of the exchange of matter takes place within, the outer surface is the main path of energy flux to and from the leaf. This separation of the sites of exchange of energy and of matter is probably the most important single consequence of the evolution of the chlorenchyma-epidermis-stoma type of construction ubiquitous in the photosynthetic organs of land plants, which has as good a claim as the tracheid to be taken as a significant diagnostic feature for taxonomic purposes. It only lacks a name.

The problem is to understand what difference it makes to behaviour in an aerial environment that a leaf should have this structure instead of a simpler one like mosses for example. Common opinion at present seems to be that it is something to do with water conservation, and an extension of this view might be that the separation of the sites of energy and matter exchange makes possible a separate control of these fluxes. It may be found that any such control is incomplete, but this would not eliminate the advantage such a system would have. A proper understanding of the way a leaf functions as a machine for living on atmospheric carbon dioxide and light necessitates a systematic investigation of all the energy and material exchanges and the literature shows that this is well under way (Gates, Keegan, Schleter & Weidner 1965, Evans 1963, Tibbals, Carr, Gates & Kreith 1964).

Energy and material exchange

The main energy exchange between a plant and its environment takes place as a result of radiation and convection, the latter being a special case of conduction. The laws governing these two primary physical processes are known, but in the case of convection their application to plants is difficult because of the irregular shapes involved. However, the rates of both processes are determined in part by the difference between the temperature of the object and that of the surroundings. The matter exchanged consists of water, oxygen and carbon dioxide. The dependence of nett assimilation rate on leaf temperature is commonplace, and it may be necessary to estimate absolute tissue temperature to better than 1°C. The computation of heat exchange by natural and forced convection has been dealt with by Gates (1962). Transpiration is the more difficult matter and it is only possible here to be concerned with the water vapour pressure (or density) deficit between the wet surface in the leaf and the air, which is the physical origin of the water loss, and which must be determined from a knowledge of the temperature of the evaporative surface, the air temperature, the relative humidity and tables of saturated water vapour pressures at different temperatures. We have only a few ways of measuring temperatures, and it is necessary to select the most suitable in each case. Only the use of the thermocouple, the thermistor and the radiometer will be considered here.

Radiative energy exchange

Energy is emitted by any object whose temperature is above absolute zero, in proportion to the fourth power of its absolute temperature. The only factor modifying this energy is a property of the surface of the object known as its emissivity. This property is inversely related to reflectivity and only when the emissivity takes the maximum value of 1 will the maximum energy be radiated. The energy J leaving a perfect radiator may be computed from the formula

$$J=\sigma T^4,$$

where σ is a constant whose value determines the units of J. In the case of an imperfect radiator whose surface emissivity is less than unity, J is diminished such that

$$J=e\sigma T^4.$$

e is numerically equal to absorbtivity, since bodies which are good emitters are also good absorbers at any given wavelength.

This energy flux occurs in all parts of the spectrum, but is unevenly distributed so that there is a wavelength at which the flux has a maximum value. If this wavelength is λ_{max}, then less than 1 per cent of the total energy is emitted at wavelengths less than $0{\cdot}5\,\lambda_{max}$, and less than 5 per cent is emitted at wavelengths greater than $5\,\lambda_{max}$. For the sun λ_{max} is at about $0{\cdot}5\,\mu$, in the visible part of the spectrum ($0{\cdot}4$–$0{\cdot}7\,\mu$). For bodies near room temperature (20°C) the maximum emission occurs at wavelengths around $10\,\mu$, so the effective limits for this 'thermal' radiation flux are from 5–$50\,\mu$. A leaf radiates in this part of the spectrum, while receiving radiation from the sun in a band of much shorter wavelengths ($0{\cdot}25$–$2{\cdot}5\,\mu$ if the figure above are followed, but convention places the limits for experimental work at $0{\cdot}3$–$3\,\mu$, which corresponds with the transmission of glass). It is thus possible and convenient to treat these two energy fluxes separately.

Consider a leaf in a simple environment consisting of soil below and sky above, at night.

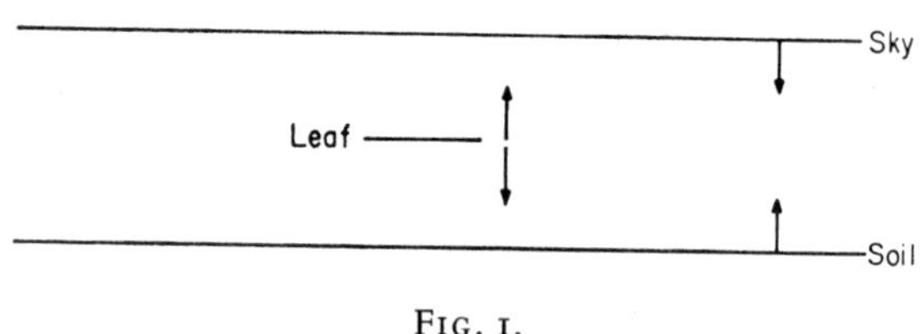

FIG. 1.

The arrows indicate the thermal radiation fluxes due to the three components.

The upper surface of the leaf sees only sky, and the lower surface sees only soil. These are the conditions in which the energy exchange for the two surfaces may be treated in the simplest way. If the temperatures of the three components of the model are known, the energy fluxes can be found from tables or by computation. But we are unable to apply thermocouples to the sky, and errors are involved in measuring the temperatures of rough stuff like the soil or delicate objects of low heat capacity like the leaf. However, we have instruments which can measure the radiant energy flux from objects directly; the solarimeter is a special type of radiometer, in which the sensitivity is limited to a selected band of wavelengths characteristic of the sun by covers of glass. A radiometer pointed at the sky or the leaf or the soil will measure the radiant energy flux from them. From this the temperature of the surface may be found, provided that the emissivity is known, and provided that the observer is prepared to treat the object as if it had a surface. In the case of the soil, for example, the radiometer will receive energy from the top of the lumps and from deep in the cracks, places whose temperature may be very different. The radiometer will give an indication of some mean temperature, and the soil will be behaving to the radiometer as if it had that temperature. This method of measuring radiant flux, therefore, gives an apparent surface temperature, which may or may not be the real one. In the case of the soil, there is no real surface temperature. Nor can the sky have a temperature, but the matter constituting the atmosphere radiates energy down to the leaf as well as out to space, and so far as this flux is concerned behaves as if it has a certain temperature. So it may be useful to speak of apparent sky temperature. The apparent surface temperature of the leaf might be supposed to refer to a more definite plane, but in this case the situation is complicated by the optical properties of the material of which the leaf is made.

Suppose that a radiometer is looking at a leaf and that the leaf is made entirely of water. The water is emitting energy by virtue of its temperature, but since the water is somewhat transparent to its own emitted infra-red flux, the energy reaching the radiometer comes from a surface layer of water of some appreciable thickness. Emissivity in a transparent medium is a volume and not a surface property. If the radiometer sees mainly in the 7–15 μ waveband, then this surface layer is about 20 μ thick. Water deeper than this will contribute a negligible amount of radiation to the outgoing flux. The presence of organic material in the leaf will reduce the effective surface thickness, but I do not know how large this effect will be. If there is a temperature gradient across the effective surface layer then the temperature indicated by the radiometer will be different from the surface temperature and nearer to that inside the leaf. If the radiometer is made to see only those wavelengths poorly absorbed by water then the effective surface layer will be increased in thickness, and it may be technically possible to use this effect to arrive at a more accurate measure of the interior temperature of a leaf. McAlister (1964) has shown how this principle may be employed to measure the temperature of ocean water at two different depths by radiometry, and gives further details of the physical principles involved.

Thus, when radiometric measurements are made in order to find the temperature of the source, there are three sources of ambiguity. First the emissivity of the surface may not be unity, so that the temperature estimate ought properly to be called an equivalent emissive temperature. Secondly, the source may be opaque but so rough that there is a range of real temperatures involved. Third, if the source is at all transparent in the long wave region of the spectrum, the estimated temperature refers to a layer of material under the true surface. Since biological situations usually involve the latter two conditions, it seems reasonable to continue to use a term such as apparent surface temperature, even though it is itself ambiguous. Equivalent emissive temperature will do as well, but appears more precise than may be warranted in some cases.

To return to the leaf and environment model of Fig. 1.

Since it is possible to measure the flux pattern directly, it is not necessary to compute it from the temperatures of the components. But we have seen that if the model is to be extended to a more natural situation involving convective energy exchange, and if the rate of water evaporation is to be properly related to environmental conditions, then it becomes essential to know the temperature of the air and the leaf in absolute units.

Air temperature

Radiometric determination of apparent sky temperature gives a figure which applies to a layer of the atmosphere that may be many thousands of feet thick. The air temperature near the radiometer will play a negligible role in this measurement. The usual method of measuring air temperature uses a thermocouple or a bead thermistor. The smaller these elements are the more closely they are coupled to air temperature, and the less they are affected by incident sunlight, which by warming the sensitive element causes an over-estimate to be obtained. Fortunately, the heating produced by incident radiation on thermocouple wires and thermistor beads can now be estimated from the tables published by Duchon (1963, 1964) and the errors found. Radiative heat loss to a clear sky at night may cause an underestimate of air temperature. When the element is shielded to avoid this source of error, it is necessary to be sure that convection is still sufficient.

Surface temperature

Radiometric estimation of apparent surface temperature gives more information about the bulk of the tissue than its actual surface, but since in practice the radiometric surface layer may be thinner than 60 μ this temperature may be very near to that of the surface, especially when the temperature gradient from inside to outside the leaf is low. This will be the case when solar insolation is low or absent. The radiometric temperature will not be affected much by the temperature of the intervening air, unless the humidity is high. There may then be enough water between the radiometer and the leaf to begin to contribute to the energy flux between the two, as well as acting as a selective filter. It is possible to estimate this error which has been described and quantified by Mitchell (1965), who points out that some types of radiometer using cooled detectors may be more affected by this source of error than those operating at ambient temperatures. The effect is easily detected or demonstrated by breathing into the radiation path.

Thermocouples or thermistors placed on the surface take information from the surface by conduction and radiation, and from the air by convection. They can never give the right reading for either temperature unless there is no temperature gradient, and that is the least interesting biological case. The usual situation is that the surface is warmer than the air because of heat being generated within. Apposed thermocouples then read too low. Radiometry will give a higher figure, but if the radiometric apparent surface is part of the temperature gradient then this figure may be too high. One can conclude that the surface temperature required for convection calculations lies between the values obtained by thermocouples and by radiometry, while if leaf transpiration is the matter of interest radiometry will give the best measurement because it gives more information on the interior conditions. It ought to be possible to combine the two methods into a procedure for extrapolation of the interior temperature of standard biological materials, as two points on a gradient are

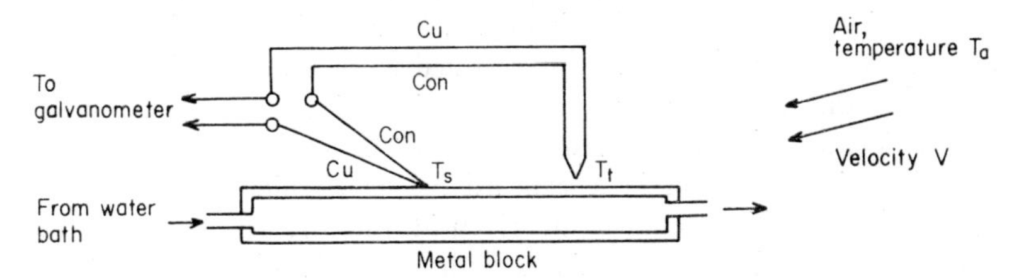

FIG. 2. Thermocouple test bed. A copper-constantan thermocouple is shown arranged so that one junction assumes T_s the temperature of the surface of the water filled metal block, and the other assumes a temperature T_t dependent upon the degree of coupling to air temperature T_a. The latter junction is the experimental one whose performance is being tested.

being obtained. Although the smaller the physical dimensions of the thermocouples or thermistors the more they couple to air temperature, when measuring surface temperatures this effect is outweighed by the fact that the small element becomes more effectively buried in the quiet boundary layer close to the surface, and therefore becomes uncoupled from the turbulent distant air. The almost instinctive practice of using the smallest elements practicable on the grounds of interfering least with the energy and material fluxes, or of not plugging up the stomata, or avoiding the conduction of heat to or from the selected spot, may give good results for this reason alone. The insolation error already dealt with applies in this case too, but will be diminished because of the conductive coupling of thermocouple temperature to surface temperature.

A reasonably simple test bed for thermocouples and thermistors may be set up in order to test various construction techniques and reproducability of results. In Fig. 2 the metal block is hollow and is filled with water circulating from a water bath. Air is blown across the surface of the block at a known velocity. It is usually easy to arrange that the block is about 10°C cooler than the air. A thermocouple pair is soldered or welded into the block such that the junction is flush with the surface. The experimental thermocouple pair is connected to this master pair so that they are opposed and the nett output measures the difference of their temperatures. If the floating pair is nowhere near the surface then T surface$-T$ air will be indicated. When the experimental pair is held onto the surface in whatever way is being adopted as standard, a temperature nearer the surface value will be indicated, with the best possible reading being zero, or T thermocouple$=T$ surface. The ratio of $Ts-Tt$ to $Ts-T$ air gives an arbitrary coupling factor which indicates how much information is taken from the air stream. Ideally this factor should be as low as possible, and constant for repeated applications of the thermocouple to the surface. In the case of thermistors it is necessary to choose two of identical characteristics and to embed one on the surface, connecting them in opposition in a standard bridge circuit. Some specimen results for two thermocouples are given in Table 1. The surface of the block was smooth but was painted with 'Britannia' black heat-resisting finish type D (by Roberts Ingham Clark & Co. Ltd) in order to simulate the high long wave infra-red emissivity of natural surfaces. The most spectacular improvement is obtained by the use of silicone oil as a thermal adhesive, but the biologist who does not wish to modify the surface in any way may take advantage of the useful effect of blackening the thermocouple wires and junction. The preferred use of thin wires and leading them along the surface isotherm is well known.

TABLE 1. Coupling factors for 2 thermocouples of differing wire thickness under different conditions of application to a standard surface.

Conditions	Wire diameter	
	0·5 mm	0·12 mm
(1) Thermocouple tip only in contact with surface of block.	0·8	0·6
(2) Wires laid flat along surface for 3 mm behind the junction.	0·4	0·27
(3) Wires as (2) above, blackened with paint of high emissivity in the infra-red.	0·34	0·14
(4) Wires as (3) above, dipped in silicone oil before placing on the surface.	0·23	0·05

Radiometers

Dogniaux (1963) has listed a variety of instruments including various types of radiometer, and gives a definition of each in terms of what it measures. He also gives a most useful table of conversion factors for the various units of energy, power, intensity and luminous flux encountered in this type of work. Gates (1962) described a variety of instruments for measuring radiation flux, but since then some new instruments have become available. These have the advantage of an optical system which enables the operator to focus the instrument on a particular area whose temperature is to be estimated. If the emissivity is known, the required correction can be applied automatically. It has also become easier in the last year or two for those with some knowledge of electronics to use cheap ready made amplifiers of small dimensions which operate from low voltage batteries. This makes the whole problem of measuring small voltages in the field much easier. The experimenter is supplied with an amplifier in a small box with a few connecting pins, and instructions from the manufacturers (e.g. Anon a and b) on the connections to be made for a variety of purposes. The main difficulty for our applications is the effect of varying ambient temperature on the amplifier characteristics. This may be overcome by building the entire electronic system inside a thermostatted block. Since the components are all small, the block does not have to be very big and the energy required may be obtained from rechargeable batteries. The entire system of amplifier plus thermostatted block plus batteries need weigh only a few pounds, and represents an easy way of obtaining information from thermocouples, thermistors, solorimeters and even a crude radiometer in field conditions.

A radiometer for field use

The basic radiation detector for the following instrument is a thermopile mounted at the apex of an effectively conical aperture within a block of aluminium. Radiation from outside falls on the thermopile, and as the thermopile itself is a nearly perfect black surface, it also radiates out through the opening. The voltage generated by the thermopile is proportional to the difference in the incoming and the outgoing fluxes, and so the output is affected by thermopile temperature as well as by the intensity of the incoming flux. To eliminate this variable the block containing the thermopile is thermostatted at 30°C, and it also contains the amplifier and associated components necessary to allow a robust meter to be used to indicate the output of the thermopile. Fig. 3 show the arrangement. The filter (F) limits the incoming (and out-going) radiation to a band between 5 and 9 μ, so that the radiometer does not see radiation from the sun when it is reflected off the leaf or object under study. Some response is obtained if the radiometer is pointed at the sun, so apparent sky temperatures in the solid angle of about 60° including the sun can not be measured.

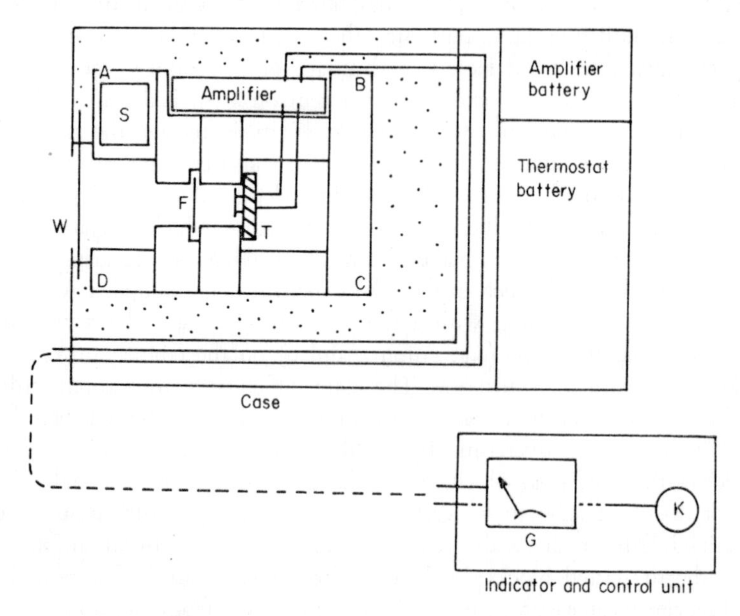

FIG. 3. A field Radiometer. Together the 2 units weigh about 3 kg. ABCD is an aluminium block 6·5 × 5 × 4 cm cut into several sections for ease of construction. The thermopile *T* is exposed to the exterior through the filter *F* and the thin polythene window *W*. The amplified output from the thermopile is displayed on the meter G, and there is a zero set control K. Just in front of the filter there is a shutter which is not shown in the diagram, and which is kept shut at all times except when a reading is being taken. The block is controlled at a temperature of 30°C ± 0·1°C by means of a miniature mercury-in-glass contact thermometer and solid state relay in the space S and heaters distributed throughout the block. The block is insulated with expanded polystyrene, shown stippled. The amplifier is a Fenlow type AD2000, which takes up to 20 mA from two 15 V batteries. The thermostat takes an intermittent current of 0·7 A from a 6 V battery. With the gain of the amplifier fixed at a value suitable for field use an input between 0·34 and 0·76 cal/cm^2/min can be measured to 1 part in 40. At maximum gain compatible with stability the span is reduced to 0·16 cal/cm^2/min. The meter readings are linear with energy input but not with object temperature.

The radiometer responds to radiation from all matter within this cone of view, so that measurements on a single leaf can only be made if the leaf is large and quite close to the aperture. Otherwise an integrated measurement is made over the vegetation canopy or the area of soil at which the instrument is directed. This instrument was originally built so that the radiation flux present in controlled environment rooms could be measured and comparisons made from room to room and with outdoor conditions. Accuracy better than 1 °C apparent surface temperature was not necessary. The instrument was operated in the field during a teaching excursion to the Cairngorm mountains, and was fitted for that occasion with alternative input terminals so that a solarimeter and thermistor bridge could be used to measure solar radiation and various temperatures. The batteries were recharged nightly.

The total cost of the instrument was £70. Full details of construction are not given here because this is a field of technology that is evolving at an ever increasing rate and more efficient designs are now possible. Full details of my instrument are, however, available from me on application.

REFERENCES

ANON (a) *The Applications of Linear Microcircuits.* S.G.S.Fairchild Ltd.

ANON (b) *Notes on the Applications and Use of Operational Amplifiers.* Fenlow Electronics Ltd.

DOGNIAUX R. (1963) Terminologie des grandeurs et instruments de mesure du rayonnement en meterologie. *Arch. Met. Geoph. Biokl.* B. **12**, 224–40.

DUCHON C.E. (1963) The infra-red radiation temperature correction for spherical temperature sensors. *J. Appl. Meteorol.* **2**, 298–305.

DUCHON C.E. (1964) Estimate of the infra-red radiation temperature correction for cylindrical temperature sensors. *J. Appl. Meteorol.* **3**, 327–55.

EVANS L.T. (1963) Editor; *Environmental Control of Plant Growth.* New York and London.

GATES D.M. (1962) *Energy Exchange in the Biosphere.* New York.

GATES D.M., KEEGAN H.J., SCHLETER J.C. & WEIDNER V.R. (1965) Spectral properties of plants. *Appl. Opt.* **4**, 11–20.

MCALISTER E.D. (1964) Infra red–optical techniques applied to Oceanography. *Appl. Opt.* **3**, 609–12.

MITCHELL D. (1965) Effects of water vapour in Biothermal Radiometry. *Life Sci.* **4**, 1267–74.

TIBBALS E.C., CARR E.K., GATES D.M. & KREITH F. (1964) Radiation and convection in conifers. *Amer. Jour. Bot.* **51**, 529–38.

THE MEASUREMENT OF CLIMATE IN STUDIES OF SOIL AND LITTER ANIMALS

AMYAN MACFADYEN

Department of Zoology, University College, Swansea, Glam., Wales
and
Molslaboratoriet, Femmøller, Denmark*

SUMMARY An outline is given of the special needs and problems encountered by the soil ecologist in measuring environmental factors; difficulties include extreme climatic gradients and heterogeneity of physical properties and of living populations. Two main fields of ecological work, involving the study of 'behaviour' and of 'performance', are recognized. Although the latter may demand elaborate technical equipment it is sometimes possible to avoid this by, for example, relating metabolic rates in the field to those measured under known laboratory conditions, or by using simple integrating devices. Examples deal with the measurement of temperature and of carbon dioxide.

INTRODUCTION

The information about the physical environment which an ecologist requires depends entirely on the kinds of organisms he is working with and which physical attributes he is interested in. For instance, the animal ecologist's idea of preferred levels of temperature, humidity etc. derives from the ability of mobile animals to move to and actively select certain regions of the environment and to avoid others. It is often assumed, but not often demonstrated, that the 'preferred' regions of the habitat are in some way optimal for metabolic activity, growth, survival, reproductive activity or other biological measurements of performance. The plant ecologist, on the other hand, has little use for the concept of preferenda because, of course, most plants are not mobile. Plants do, however, show a plasticity in growth and reproductive performance which is astonishing by comparison with most animals, and different features of

* Present address: The New University of Ulster, Coleraine, Northern Ireland.

growth, reproduction, survival and so on frequently have optima at somewhat different combinations of environmental conditions (Harper 1961). Both plant and animal ecologists, therefore, have reason to measure the spatial and temporal variation of physical factors throughout a habitat. The simplest kind of problem is probably encountered by students of sessile organisms when they define the conditions which prevent survival or success. It is then only necessary to determine those places or occasions at which extremes of temperature, soil moisture etc. occur or to measure the extremes which are encountered at a particular position. The most difficult to study are the small, mobile creatures which move freely through a habitat in which there are steep physical gradients. These difficulties are greater when it is necessary to know, not just the physical extremes encountered, but also diurnal and seasonal succession of such factors. In the study of soil and litter animals we encounter all these difficulties in their most acute form and I must admit at the outset that we are very far from overcoming them all. The paper which follows consists of some notes and reflections on my own experiences and difficulties in this field.

The outstanding climatic characteristic of the upper layers of most soils is the occurrence of very steep gradients of soil temperature and of dependent factors such as moisture. In exposed soils, even in temperate latitudes, differences of 10°C and more between the surface and the soil at 1 cm depth are common. Diurnal changes at the surface itself often exceed 40°C in clear weather and, even at 5 cm depth, diurnal variations of 20°C are usual. The great majority of soil-dwelling animals occur at depths less than 5 cm, and are thus exposed daily to temperature changes which are greater than those usually encountered in Stephenson screens at opposite ends of the continent of Europe, or than the differences between average summer and winter air temperatures in the same place. This extreme spatial and temporal variability and the consequent temperature gradients constitute the greatest obstacle to accurate work in this field. They may help to explain why we are sometimes happy to sacrifice accuracy to realism: to dispense with readings in decimals of degrees if only we can approach a little way towards measuring the conditions actually experienced by the organisms under study.

Further, it will be appreciated that, if for any particular study we can define precisely what we must know, leave out un-essential information, and adapt the measurement techniques to the problem, there is a better chance of gaining essential ecological information than if we set out to measure everything to the highest possible degree of accuracy.

Methods for 'behavioural' studies

In work on soil animals (and indeed in most fields of animal ecology) there are two rather distinct types of study for which knowledge of physical quantities is required. The first is concerned with accounting for the observed preferenda and tolerance ranges of organisms and with linking field behaviour to the results of laboratory experiments. In work of this kind a series of short-term measurements or recordings, possibly repeated on several occasions and made in close proximity to the animals, is usually required. It is often possible to obtain this information with relatively simple 'spot reading' equipment if an animal such as a large insect or other arthropod can be located and its distribution defined.

Most satisfactory results of this kind have been obtained by ecologists (e.g. Nørgaard 1945, 1951, Linde & Wondenberg 1951) working with surface-living animals such as spiders and beetles; in such work it is necessary to account for interactions between different factors, both physical and physiological (for example, changes in behaviour with hunger, water shortage, reproductive condition, etc.); when this has been done, the results are frequently clear-cut and consistent. When working with the numerous small invertebrates inhabiting the deeper soil layers, it is difficult to follow individuals with temperature and humidity probes, but Berthet's (1964) successful tracing of oribatid mites using radioactive paint indicates the possibility of doing so. The usual tendency with such animals has been to resort to sampling: many probes at known depths define the mean and the error for the temperature or other factor and the fauna is extracted from numerous samples in the different layers of the soil. The results from this procedure are usually inconclusive, chiefly, I suppose, because both the physical properties of the soil and the distribution of the fauna are extremely patchy and it is very difficult to obtain valid differences by this 'sledge hammer' technique.

Such work, in which an attempt is made to relate field distribution to behavioural factors is, on the whole, relatively undemanding as regards microclimatic techniques but is very exacting upon the patience of the observer and his skill as a naturalist. On the technical side, such devices as thermistors and thermocouples are usually adequate for temperature measurements, if intelligently used and if account is taken of the steep gradients which occur. Humidity can frequently be measured with Solomon's (1945) cobalt thiocyanate papers or Nielsen & Thamdrup's (1939) sulphuric acid tubes. Simple Piche evaporimeters have their uses in more open environments.

Methods for 'performance' studies

The second type of study is concerned with the performance of animals throughout their life cycles. By performance I mean physiological characteristics such as metabolic rate, growth rate, mortality, reproductive rate. This type of study is also concerned with the relative performances of different species, such as predators and their prey, and competitors living in the same habitat. In some cases it is imperative, for a full understanding of performance, to have continuous measurement of the climate actually encountered by the animal. No good examples appear to be available for soil animals but I am thinking of studies such as Messenger's (1964) on the rate of increase of an aphid and its hymenopteran parasite under different climatic regimes. By the use of variable climate chambers, programmed to simulate the conditions occurring in different parts of California, it was possible to predict the outcome of control measures in the field. A similar satisfactory relationship was found with the climate by the same workers (Force & Messenger 1965) in the case of competition between different parasite species.

In such a case it is difficult to see how we can dispense with continuous records of climatic factors over long periods. We do not know which characteristics of the climatic regime are relevant and important. However in other cases it has proved possible to obtain close correlations between, for example, metabolic rate and mean temperature in the laboratory. In order to extrapolate back to field conditions we would like to know what was the mean temperature experienced by a population over a long period. If we knew exactly where the animals occurred at all times the information could be obtained from continuous records at many positions throughout the animal's range. In the case of exposed soils with sharply stratified temperature regimes this is probably the only realistic approach by direct instrumentation. However most of us have shirked this problem and compromised by avoiding highly exposed habitats and by arguing from the known absence of marked diurnal and seasonal migration of the fauna in such sheltered regions. In this case static probes at known positions will give adequate information. At the present time nobody to my knowledge has done better than this.

Ideally one would like to saddle the animal with some miniature telemetry device, and the recent development of extremely small but powerful and simple UHF 'Gunn effect' oscillators (Bowers 1966) may well make this possible for animals down to the size of small mammals at least. Another method not yet available for very small animals is the use of miniature radiation detectors for integrating the number of hours spent in a particular habitat (French, Maza & Aschwanden 1966). Small rodents labelled with radiation detectors sutured to them were kept in a research area subjected to radiation from a source suspended above it. The proportion of time spent by the mice on the ground surface could be measured from the extent to which the crystals were activated. The most promising approach to obtaining true field metabolic data at present, however, seems to me to be the measurement of the rate of elimination of radioactive isotopes from an animal's body. Preliminary work by Reichle & Crossley (1967) and Crossley (1966) shows, for instance, that the retention of radioactivity by leaf beetles (its biological half-life) is temperature-dependent and that the animals, given radioactive food, and monitored before and after liberation in the field, can be used as mobile biological thermometers in this way. There are, however, complications particularly dependent upon which isotopes are used and how they are dispersed through the body.

These methods appear to me to be promising for the future but have not yet been applied to soil invertebrates. Until such methods are available, therefore, we must avoid the most technically exacting environments and content ourselves with a static measurement of relatively immobile populations.

Devices for integrating temperature

I would suggest that there are many occasions when it is more important to obtain replicated measurements of mean temperatures than to employ much more expensive continuous recorders and to restrict the number of recording stations. This applies particularly to soil work because, in such a heterogenous medium, it is very difficult to define the biologically significant layers precisely and to account for variations in depth and structure of the overlying litter and vegetation. Thus replicated reading points are an essential safeguard against placing too great a reliance on an anomalous reading position. Although Loggers are now available which can accept data from a number of transducers and their records can be analysed to provide mean and excursions, I suggest that the relative costs of such equipment should be carefully weighed against those of simpler devices.

A number of integrating devices have been described in the literature. These include a mechanical recorder due to Dahl (1949*a*, 1949*b*) which has the advantage of providing a frequency distribution of temperature recurrences but is too large for work in thermally stratified soils, the sucrose inversion type, the latest version of which is due to Berthet (1960), and types employing a thermistor in series with an electrolytic voltameter (Macfadyen 1956). In the inversion type a rather large glass tube of sucrose solution is exposed in the site and the extent of inversion is measured with a polarimeter; inversion is exponentially related to temperature with a Q_{10} of about 2. It thus roughly parallels the metabolism/temperature relationships of many organisms. Apart from the initial cost of the polarimeter, the method is cheap and well adapted to making very large numbers of measurements in sites such as woodland.

The thermistor/electrolytic integrator also has an exponential law and the exponent can be modified if desired (Macfadyen 1949). The probe itself is very much smaller and it is therefore more suitable for exposed sites than the sucrose method. The original version employed a silver voltameter and this involved transport of rather fragile electrodes to the laboratory for washing, drying and weighing: the use of silver nitrate is also rather unsatisfactory for aesthetic reasons. Nevertheless two long-term studies have been completed with this device (see also Healey 1965) in which the effects of season, soil depth, aspect and other factors were clarified. The apparatus is relatively cheap (about £1 per station); the chief source of lost readings has proved to be sabotage by humans and small mammals; both these are avoidable in a small device which is so easily concealed.

An improvement in the thermistor type integrator which, however, at present increases the cost to about £5 per station, can be made by using a mercury integrator (Macfadyen & Webb, in preparation). The most reliable version of this is the Curtis meter (Model 190 unmounted from Curtis Instruments Inc., 351 Lexinton Avenue, Mount Kisco, New York). It consists of a 35 mm precision capillary tube into which platinum electrodes are sealed at the ends. The tube is filled with mercury except for a small gap containing electrolyte. With the passage of current the gap migrates along the tube and the movement can be measured with a Vernier caliper. When the gap has moved to one end the meter is reversed. Ruben-Mallory alkali cells are used as a source of current and with a current of 1 mA the gap moves about 2·5 mm in 10 hr. Thus the normal F23 thermistor (resistance about 1000 ohms at 20°C) and a single cell are suitable for daily readings but the F15 with several cells in series is better for weekly readings. The method has the great advantage that all reading is done in the field. Except for the fact that the Curtis meters need careful handling and must not be overloaded or driven beyond the end of their range, no serious practical difficulties have been encountered in the use of this integrator. We started with the use of an English version of the Curtis meter but have now decided that the American ones are cheaper, more reliable and more readily obtainable.

I have concentrated on temperature measurement and its recording and integration because temperature is the quantity which I have most frequently needed to measure myself. Also this provides an opportunity to illustrate the theme of suiting the method of measurement to the ecological information required. I shall now briefly mention another field in which I have been able to use simple equipment in a similar manner.

Measurement of soil carbon dioxide

There is a good deal of evidence that carbon dioxide is important to soil organisms, both because they are attracted or repelled by gradients of the gas (e.g. Klinger 1958) and through its effect on their respiratory activity. The activity of parasitic trombid nymphs is stimulated by carbon dioxide originating from potential vertebrate hosts and Collembola have been shown to follow a carbon dioxide gradient to the vicinity of plant roots. On the other hand some soil fungi have been shown to be surprisingly sensitive to levels of carbon dioxide which quite frequently occur in soil both in culture (Burges & Fenton 1953) and in the soil itself (Macfadyen, in preparation). It is therefore of considerable interest to determine what levels of carbon dioxide occur under natural conditions.

A simple device for this purpose described by Martin & Piggott (1965) consists of a glass tube of bicarbonate solution with a small window covered with PTFE (polytetrafluoroethylene). The solution equilibrates with the surrounding soil carbon dioxide and this can be calculated from the pH of the buffer solution. Unfortunately the device described by these authors takes more than a week to reach equilibrium and I have found it more convenient to use simple thin-walled polyethylene 'sausages' about 1 cm × 15 cm. These are made with the aid of a heat sealer, and the ends are cemented first to polyethylene tube and then to PVC (polyvinyl chloride) tube with Araldite. This can be stoppered with a tapered glass plug. The contents are sampled with a syringe. An alternative use of similar 'sausages' is to inflate them with air and to sample the air by means of a simple gas analysis apparatus. Full details of construction and performance will be published elsewhere (Macfadyen, in preparation).

The above are some examples of the kinds of problems which have been encountered by one ecologist when trying to measure environmental factors in soil. I have made no attempt at complete coverage of all kinds of factors because that is a function for our whole symposium. I hope however that my paper will serve to encourage other biologists to attack physical and chemical problems directly, improvizing new methods where necessary rather than accepting existing techniques regardless of their relevance to the highly specialized demands which are made by so many of our habitats. I hope too that it will stir the indignation of colleagues in the physical sciences to suggesting better and more accurate methods for us to use.

REFERENCES

BERTHET P. (1960) La mesure écologique de la température par détermination de la vitesse d'inversion due saccharose. *Vegetatio* **IX**, 197–207.

BERTHET P. (1964) Field study of the mobility of Oribatei (Acari) using radioactive tagging. *J. Anim. Ecol.* **33**, 443–9.

BOWERS R. (1966) A solid-state source of microwaves. *Sci. Amer.* **215** (2), 22–31.

BURGES A. & FENTON E. (1953) The effect of carbon dioxide on the growth of certain soil fungi. *Trans. Brit. mycol. Soc.* **36**, 104–8.

CROSSLEY D.A. (1966) Radioisotope measurement of food consumption by a leaf beetle species, *Chrysomela knabi* Brown. *Ecology* **47**, 1–8.

DAHL E. (1949*a*) A new apparatus for recording ecologic and climatic factors. *Science* **110**, 506–7.

DAHL E. (1949*b*) A new apparatus for recording ecological and climatological factors, especially temperatures over long periods. *Physiol. Plant.* **2**, 272–86.

FORCE D.C. & MESSENGER P.S. (1965) Laboratory studies in competition among three parasites of the spotted alfalfa aphid *Therioaphis maculata* (Buckton). *Ecology* **46**, 853–59.

FRENCH N.R., MAZA B.G. & ASCHWANDEN A.P. (1966) Periodicity of desert rodent activity. *Science* **154**, 1194–5.

HARPER J.L. (1961) Approaches to the study of plant competition *Symp. Soc. exp. Biol.* **XV**, 1–39.

HEALEY I.N. (1965) *Studies on the Production Biology of Soil Collembola*, Ph.D. Thesis, University College of Swansea.

KLINGER J. (1958) Die Bedeutung der Kohlendioxid-Ausscheidung der Wurzeln für die Orientierung der Larven von *Otiorrhynchus sulcatus* F. und anderer bodenbewohnendes phytophager Insektenarten. *Mitt. Schweiz. ent. Ges.* **31**, 305–269.

LINDE R.J. VAN DER & WONDENBERG J.P.M. (1951) On the microclimatic properties of sheltered areas. *Meded. Inst. Toegep. biol. onderz. Nat.* **10**, 151 pp.

MACFADYEN A. (1949) A simple device for recording mean temperatures in confined spaces. *Nature, Lond.* **164**, 965.

MACFADYEN A. (1956) The use of a temperature integrator in the study of soil temperature. *Oikos* **7**, 56–81.

MARTIN M.H. & PIGOTT C.D. (1965) A simple method for measuring carbon dioxide in soils. *J. Ecol.* **53**, 153–5.

MESSENGER P.S. (1964) Use of life tables in a bioclimatic study of an experimental aphid-braconid wasp host-parasite system. *Ecology* **45**, 119–31.

NIELSEN E. TETENS & THAMDRUP H.M. (1939) Ein Hygrometer für mikroklimatologische Untersuchungen. *Bioklim. Beibl.* **6**, 180–4.

NØRGAARD E. (1945) Økologiske Undersøgelser over nogle danske Jagtedderkopper. *Flora og Fauna* **51**, 1–38.

NØRGAARD E. (1951) On the ecology of two Lycosid spiders (*Pirata piraticus* and *Lycosa pullata*) from a Danish sphagnum bog. *Oikos* **3**, 1–21.

REICHLE D.E. & CROSSLEY D.A. (1967) Investigations on heterotrophic productivity in forest insect communities, *in*: Petrusewicz D.K. (Ed.) *Principles and Methods for the Study of Secondary Productivity*, *Ekologia Polska*, (I.B.P. Working Meeting, Warsaw, 1966).

SOLOMON M.E. (1945) The use of cobalt salts as indicators of humidity and moisture. *Ann. appl. Biol.* **32**, 75–85.

THE MEASUREMENT OF WIND SPEED AND DIRECTION IN ECOLOGICAL STUDIES

J.M.Caborn

Department of Forestry and Natural Resources, University of Edinburgh

SUMMARY An attempt is made to survey workable methods for the measurement of wind speed and direction, from the point of view of the ecologist who may wish to measure these factors of the environment. Attention is drawn particularly to the general range of equipment available and the operating principles involved. Various modifications of existing instruments and new designs, applied successfully in field studies of wind conditions, are mentioned briefly in order to suggest further possibilities.

The introduction to the paper deals with the objectives of the study and, correspondingly, the type of data that are applicable. The choice of parameter and the choice of sampling intensity or degree of resolution required from continuous records govern the general characteristics of the equipment necessary. However, cost/benefit considerations and practical problems encountered in the field may modify these requirements.

INTRODUCTION

The purpose of this paper is not to review recent advances in, or specialized techniques for, the measurement of wind speed and direction but to survey some established methods in relation to the needs of the ecologist, who measures, or wishes to measure, environmental factors 'as part of a wider research programme'.

It is precisely this phrase which is a matter for concern in ecology. As Gates (1962) has said, ecologists have in the past been guilty of measuring certain climatological factors as interesting features of the environment, without relating these parameters to physical processes affecting the organism in question. If wind data can be exempted from this criticism more than certain other parameters, it is perhaps because the anemometer has been less readily available than, say, the thermometer. The fact that statistical correlations between wind speed and biological observations have often been forthcoming does not of itself alter the situation. In Hull's words (1959), 'the validity of an argument does not guarantee the truth of the conclusion'.

The experimental ecologist contemplating the measurement of wind should first satisfy himself that there is a good answer to the question: why wind is to be measured. If he can also satisfy the physicist or the meterorologist, so much the better; hypotheses which may sound convincing from an ecological point of view may still lack a physical foundation. Such an approach is also likely to improve the potential contribution of the investigation to ecological knowledge. Whereas a purely empirical approach to an environmental problem in field ecology can rarely lead to generalizations being made from limited experimental data, a more analytical and physical approach frequently does. For instance, where the wind is studied as a separate phenomenon and thereafter related directly to some measure of biological response, without any intermediate attempt to relate the wind data to the physical processes which determine both natural and artificial environments, the ecological picture can easily become distorted.

Unfortunately, it is still all too true that ecologists often forget the basic definition of wind, which is simply air in motion: air possessing properties of density, viscosity and specific heat and the vehicle for turbulent transfer of momentum, heat, water vapour, gases, pollutants, pollen, spores and seeds.

THE DATA REQUIRED

The type, detail and duration of wind speed and direction data most applicable to the situation being explored are generally suggested by the objectives of study. For wind speed, various parameters are possible. These range from run-of-wind over a period, which is cheap to obtain, to diurnal or annual patterns which are expensive to obtain because they require some form of continuous recording. These diurnal or annual patterns may be based on daily, monthly or annual means obtained with simple instruments; or where continuous recording is possible on peak velocities such as the highest hourly wind or the difference between maximum and minimum wind speeds (the gust range); or the duration of winds exceeding a selected speed. For wind direction a cumulative measure of the relative distribution by compass points, quadrants or octants is comparatively simply obtained. In the measurement of both wind speed and direction it is the recording of variations in wind conditions with time which requires elaborate and expensive instrumentation. However, the use of self-registering instruments is nearly essential in the usual experiment where simultaneous observations are necessary at two or more sites.

The sampling intensity and degree of resolution required from continuous records are likewise questions which must be determined for each individual investigation, bearing in mind, perhaps, the problems associated with manual extraction of data from even a few weeks of wind speed and direction traces. The degree of accuracy required, or even possible, is also important; it is not uncommon to find values recorded without regard to the instrumental error. Furthermore, although cost/benefit evaluations are generally not applied deliberately at the start of an investigation (admittedly, they exercise a considerable indirect control), they merit attention. The suitability of equipment should be assessed not only on initial cost and maintenance but also on the cost of data handling in the future. Broadly speaking, the more intricate and automatic the equipment the greater is the accumulation of data, which, in order to be utilized, may warrant or demand additional expenditure on a more appropriate system of data acquisition and processing, e.g. a magnetic or punched tape output. The obvious alternative is to restrict the data collection to manageable proportions, which may be achieved by means of more modest instrumentation. On the other hand in the process of economizing on instrumentation one must not lose sight of the original objectives.

All this may seem elementary, but experience shows that it is the elementary facts which can most easily be overlooked.

GENERAL CHARACTERISTICS OF WIND-MEASURING EQUIPMENT

General characteristics of the standard range of instruments available for the measurement of wind speed and direction are given by Middleton & Spilhaus (1953) and the Handbook of Meteorological Instruments (Meteorological Office 1956). Both deal at length with the classification of instruments according to their principles of operation and also discuss basic instrumental theory.

Wind speed

Briefly, anemometers fall into three main groups:

(i) Mechanical anemometers depending upon (a) the rotation of a cup assembly (cup anemometers) or windmill or propeller (vane anemometers), and (b) the wind pressure on a suitably mounted plate or similar body (pressure plate anemometers);

(ii) Pressure-tube anemometers involving a flow of air through the instrument and dependent upon the relationship between dynamic and static pressure of the air;

(iii) Thermal and thermo-electric devices based on the cooling power of the air.

The majority of instruments regularly used in ecological studies are comprised in (i)(a) and (iii) above.

Cup anemometers

These instruments, in which the rate of rotation of the cups is directly proportional to wind speed to a sufficiently close approximation, are the best known of the mechanical anemometers. The registering mechanism may be a simple counter, giving run-of-wind in miles or kilometres in standard instruments (e.g. Met. Office Cup-Counter Mk. II) or in arbitrary units for individually calibrated instruments (e.g. Casella Sensitive Type IV Cup-Counter). Contact types, on the other hand, close a contact every 1/120, 1/60, 1/20 or 1 mile run of wind and can be incorporated in simple electrical circuits to operate a counter, buzzer or event recorder. A third type, the cup-generator anemometer, generates a.c. current which is fed to a rectifier and then to a moving coil galvanometer. Various modifications of these three registering principles have been developed, particularly during the last 10 years.

The standard cup rotor consists of three conical cups with beaded edges, although British models with three hemispherical cups (e.g. Munro Cup-Contact Anemometer) are available occasionally and current American trade literature features both 3- and 4-hemispherical-cup models, with and without beaded edges. Hemispherical cup models are suitable for use where their additional error in over-estimating fluctuating winds is within the required limits of accuracy. A recent innovation is the use by the Beckman and Whitley Co., San Carlos, California, U.S.A. of a 2-tier cup arrangement, each tier of three cups. In addition to slight variations in cup design, current models of anemometer exhibit greater variety in the ratio between cup diameter (d) and the diameter of the circle described by the cups (*D*). Middleton & Spilhaus (1953) conclude that for large instruments the ratio d/D should be less than for small cups. From a practical point of view, a heavy cup on a long arm may sag as a result of exposure to strong winds.

Cup anemometers range appreciably in size and, hence, in starting speeds; 0·25–3 mph (0·1–1·3 m/sec) in commercial models. Most overrun to some extent in fluctuating winds and at low wind speeds start to underestimate wind speed and eventually stall. Consequently, improvements in design have been directed largely towards reducing friction in the rotor spindle and in the registering mechanism, in order to improve sensitivity. Modifications in the design of the rotor, as mentioned above, are of minor importance compared to improvement in the bearings and in the inertia of the whole system. A compromise between sensitivity and ruggedness of the instrument is usually necessary.

Vane anemometers

The vanes rotate in a vertical plane and the instruments are available in several designs. Ower (1949) gives a detailed account of the theory of this instrument, which, although occasionally used in ecological studies, is more suitable for measuring steady flow. In a variable wind its error may be as much as 12½ per cent and in very turbulent conditions the vane will rotate alternately clockwise and anti-clockwise.

A new instrument apparently combining the principles of the cup anemometer and the vane is the Meteorological Office fan-type anemometer, still under development. A 6-blade propeller is held in the wind stream at the end of a direction vane, provision being made for the recording of both wind speed and direction.

Pressure plate anemometers

These are seldom used in field work. The industrial 'velometer', employed in ventilation studies, uses this principle. However, Jensen (1954) developed a large but sensitive swinging-plate instrument for his research on windbreaks on Danish heathlands. The chief difficulty in the use of these instruments is the need to take the mean of 24 successive visual readings of the deflection angle in order to obtain a reasonably accurate measure of wind speed.

Pressure-tube anemometers

Scaled-down versions of the Dines pressure-tube anemograph as used at major weather stations were also developed by Jensen (1954) but are generally too bulky for field use. Multiple pitot-static tubes, coupled to alcohol manometers, are occasionally employed for profile measurements near the ground. Woodruff, Read & Flack (1959) added directional vanes to their assemblies on 40-ft towers, one of which was mounted on a truck for traversing the field of influence of a shelterbelt. Multiple-tube manometers are best read photographically.

Thermal and thermo-electric devices

These devices for determining wind speed, the rate of cooling providing a measure of the wind flow, have been extensively developed in recent years. This applies especially to hot-wire anemometers. Simple instruments measuring 'cooling power', such as the Kata thermometer (Hill, Griffith & Flack 1916), have aroused interest from time to time for measuring low speed air flow when the air temperature is also known, but are of limited ecological importance. Hot-wire anemometers, common in wind-tunnel work, are being used increasingly in micrometeorological and environmental studies and are particularly suited to the measurement of low wind velocities near surfaces and within crops. Their theory is described by Ower (1949) and Tanner (1963); Tanner gives additional references. The principle in hot-wire anemometry is that a pure metal wire, commonly of platinum, nickel, tungsten or platinum-iridium, is heated in an electrical circuit and exposed to the air current, when is undergoes cooling. If the wire is maintained at constant temperature and, hence, constant resistance, the current varies with wind velocity. Alternatively, the current can be kept constant, the wire temperature and resistance varying with velocity.

Several commercial designs of hot-wire anemometer are available. These are generally expensive and in consequence many versions have been constructed locally for individual research projects. The portable Hastings model described by Caborn (1957) is a moderately-priced trade instrument which has proved very useful and reliable for probing the environment of small spaces. Fritschen & Shaw (1961) adapted a Dutch thermocouple-type design, which has been used in several field crop studies (e.g. Rosenberg & Allington 1964). In this instrument, the thermocouple sensors are incorporated in a 'heated bead', the heat transfer therefore being less dependent on wind direction than in the case of a wire. Long (1957) constructed a useful 'hot bulb' anemometer which employs the same principle. These anemometers are cheap to construct; they are more robust, have a more stable calibration and can be made to use a smaller heating current than the hot wire anemometer.

Sonic anemometer

In a completely separate class is the sonic anemometer (Suomi 1957) developed for turbulence studies requiring a fast response instrument. Based on the principle that sound travelling with the wind travels faster than sound going against the wind. However, because a large distance is needed between source and sensor this anemometer is likely to be restricted to micrometeorological investigations. It is mentioned solely because it illustrates that future instrument development need not be confined to the three broad categories already outlined.

Wind direction

Wind vanes and direction recording instruments are discussed in a very practical manner by Middleton & Spilhaus (1953), who illustrate simple pen arrangements for registering wind direction. An additional pen can record intervals of time. On the whole, wind direction recorders are not difficult to construct (see, for example, Schmidt & Marshall 1960) and local construction is often necessary due to the remarkable dearth of suitable instruments for field experiments on the British market. This lack is perhaps more surprising when one considers the simplicity of the circuitry involved in making a contact between a moving vane and 8, 16 or 32 compass points.

Another comparatively recent development is the use of low-torque, wire-wound potentiometers for sensing wind direction. This principle has been applied particularly in the study of wind fluctuations by means of bi-vanes, which record both azimuth and elevation angles (Yaffe 1956; MacCready 1964). Bi-vanes are available commercially in the USA but not yet in Britain.

It may be of interest to add that, although the traditional shape of wind vane still predominates, various splayed vanes and aerofoil forms designed for quick response and greater stability are on the increase in American trade literature.

Anemographs

Anemograph instruments recording both wind speed and direction are generally reserved for permanent installation at weather-reporting stations, although a Fuess anemograph operating on the cup-contact principle has been in field use at Bangor for a number of years as a control station in a long-term investigation (Gloyne 1962). The instantaneous direction is recorded at the end of each 25 km run of wind. It is doubtful, however, whether anemograph instruments are of more than limited interest to ecologists, bearing in mind their complexity, the need for mains electricity in some cases, e.g. the cup-generator anemograph, and the labour involved in chart analysis. With the development of automatic weather stations, providing data in a more convenient form, an alternative system should soon be available to British users. The new fan-type anemometer, Meteorological Office pattern, may provide the sensing element for wind in one or more of the automatic recording stations now under development by the Meteorological Office. In one version of this anemometer a dc Desynn transmission gives an output suitable for a continuous record; in another, 18 micro-switches each represent a 20° arc.

The equipment which has been mentioned above provides a general selection of that currently available and, above all, illustrates the operating principles involved. Some modifications of standard instruments and systems developed for particular ecological purposes are discussed below.

SOME EXAMPLES OF INSTRUMENTATION IN ECOLOGICAL STUDIES

Any attempt to classify wind measuring devices according to their ecological usefulness is bound to draw arbitrary and non-existent divisions. It might suggest that there is a standard ecological situation to be explored, whereas the possibilities for environmental study in ecology are infinite. Nevertheless, for those interested in the investigation of the effects of shelter against wind, Gloyne (1966) compiled a list of instruments suitable for control sites and/or semi-permanent installations and those more suited to, or developed for, special studies. This provides a useful working guide. Gloyne's 'control sites' may be regarded as similar to, if not as complete as, auxiliary climatological and crop-weather stations, details of which are given in the Observer's Handbook (Meteorological Office 1952). Whilst experimental control measurements cannot always adhere to the standard heights above ground adopted for climatological stations, it is clear that as far as possible environmental sampling should be referred to the local climatological picture.

The selection of instruments for such control stations, for long-period regional surveys of wind conditions and semi-permanent installations, where wind speed and/or direction are observed at a height of 1·25–2 m above ground, will be governed largely by the data required and by the limitations imposed by the recording equipment. Whereas micrometeorologists would appear to have concentrated their attention on the development of sensing elements, ecologists in general have been more concerned with the development of relatively simple recording equipment. Schmidt & Marshall's (1960) wind-direction recorder for remote stations and Sumner's (1965) long-period recorder for wind speed and direction, battery-operated and producing 10 records/h, are typical examples.

Battery-powered data logging systems, if not yet in common use, are becoming increasingly available for the regular monitoring of a number of variables. The Forestry Commission is understood to be using a 50-point Westinghouse system collecting data in the form of electrical analogue signals from tacho-generators (wind speed) and potentiometers (wind direction) and feeding these through an analogue-to-digital converter to a Creed punched-tape output. Sources of information can be scanned at regular intervals within the space of a few seconds. This system uses 48 V batteries, which limits the output rate as compared with mains supply. An alternative may be to use a generator. Similar systems, chiefly mains-operated, are obtainable but all are expensive (£2500 upwards). At a much lower price level (£100–£200) are small, portable systems such as the D-Mac Limpet Logger, now in use at several centres. These 1-, 4- and 10-channel units accept a variety of sensors (additional to the basic price) and convert analogue variations of voltage or resistance (with a maximum input signal of 5 V) into digital pulses for recording on magnetic tape. Translation services are available from the manufacturers. Gloyne (1966) discusses sensing equipment for the Limpet Logger.

Data-logging equipment readily lends itself to the development of sensors for wind and other factors. For crop-environment studies at Edinburgh University, Marshall (1966) modified a number of Sheppard-type sensitive counter anemometers for use with the Limpet Logger. A Ferranti toroidally-wound 357° potentiometer, fitted to one of the counter spindles, integrates wind-run over the recording interval of 15 min. The main transducer circuitry, including resistors and Mallory-type dry cells, can be remote from the anemometer. A similar potentiometer fitted to a very light vane records wind direction. A practical problem is encountered in weather-proofing the modified anemometer for long exposure in the field, without affecting air flow past the instrument. Rider (1960) found that the asymmetrical mounting of the flat-fronted contact housing on the Sheppard anemometer already affected measurement and in the latest Casella version the contact housing is spherical, following a modification adopted by Imperial College, London, research workers some time ago.

This new Casella model of sensitive cup anemometer, which supersedes the earlier counter and contact patterns, requires a 12 V supply and employs at the end of the rotor spindle a light-interruptor between a miniature lamp and photocell. Of the many modifications of the Sheppard and similar sensitive anemometers (see, for example, Deacon (1948) and Lettau & Davidson (1957)), the replacement of electrical contacts by the action of light on photo-sensitized diodes and photo-transistors has established a trend.

Other modifications are still of interest, however. Long (1957) fitted light platinum contacts to the dial of a counter-type anemometer, an alteration of particular interest because of the additional circuitry described to prevent spurious readings should the contacts chatter or remain closed as they may do in older instruments or when the anemometer stalls in low wind speeds. Middleton & Spilhaus (1953) and Tanner (1963) illustrate capacitor circuits to avoid this problem.

A similar difficulty may be encountered where a series of post office type electro-mechanical counters are employed for remote recording of wind-run from several contact anemometers (Nägeli, 1953; 1965). Such counters can be read photographically at regular intervals by means of inexpensive, fixed-focus cine cameras often available as Government surplus. In order to restrict the number of frames exposed during each closure of the contacts actuating the camera, a capacitor or thermal delay circuit can be incorporated. At Aberystwyth, Rutter (1966) has advanced this technique considerably in a system which used 20 electromagnetic counters in conjunction with a date-time clock, rotary switch and 16 mm camera, linked to a 24 V supply. Readings can be taken from negative film. An alternative to counters is to use a chart recorder. Simple event recorders are marketed in a varied price range or can be constructed from any drum-type recorder (Reigner 1964) or chart clock. In this connection, the pen arrangement mentioned for wind direction measurement is worth noting (Middleton & Spilhaus 1953).

Although few ecologists appear to have constructed their own mechanical anemometers, Nägeli in Switzerland in the early 1950's developed a cup-contact model from ping-pong balls mounted on a light spindle, which operated clock gears fitted with a platinum wire contact. The wire swept through a drop of mercury applied to a second terminal at the start of each series of observations. This anemometer is illustrated in Nägeli (1965), together with other features of anemometer and direction recorder design.

References to conventional wind direction equipment have already been made. In a different category is Baltaxe's (1962) design of a 180° tilting-plate recorder, suitable for investigations where one is concerned only with slight fluctuations about a mean direction or in selecting periods with little deviation in wind direction.

Finally, mention must be made of the many useful ecological investigations undertaken without recording equipment (e.g. Hogg 1965).

EXPOSURE OF INSTRUMENTS

The instrument manuals mentioned earlier deal with the question of instrument exposure. Further information can be obtained from the International Guide (W.M.O. 1954), Platt & Griffiths (1964) and Tanner (1963). An elementary point, not always stressed sufficiently, is that the axis of mechanical anemometers must be vertical. In addition, when mounted on arms on a vertical mast they must be far enough from the mast and other obstructions to avoid interference. Tanner (1963) and Slatyer & McIlroy (1961) give practical advice for the measurement of wind profiles. Several authors have dealt with the problems of matching pairs of individually calibrated instruments for this purpose. Where wind-tunnel facilities are not available a reasonably accurate alternative is to use a whirling arm (Middleton & Spilhaus 1953) which can be simply constructed.

CONCLUSION

In a short review it is impossible to mention more than a few of the many instrumental developments which deserve the ecologist's attention. Gloyne (1966) refers to several, for which published information is not available; some others are mentioned by Rogers & Whitbread (1966). The aim of this paper has been to indicate the breadth of equipment generally available and to present some of the ideas and principles which have led to new developments in instrumentation. In the field of environmental measurement, it is the user, faced with a practical problem to investigate, rather than the instrument manufacturer, from whom new instruments and new techniques will originate.

REFERENCES

BALTAXE R. (1962) Shelter research. *In Report on Forest Research for year ended March, 1961*, pp. 101–5. For. Comm., London.

CABORN J.M. (1957) Shelterbelts and microclimate. *Bull. For. Commn, Lond.* No. 29.

DEACON E.L. (1948) Two types of sensitive recording anemometer. *J. scient. Instrum.* **25** (2), 44–7.

FRITSCHEN L.J. & SHAW R.H. (1961) A thermocouple-type anemometer and its use. *Bull. Am. met. Soc.* **42**, 42–6.

GATES D.M. (1962) *Energy Exchange in the Biosphere.* Harper & Row, New York.

GLOYNE R.W. (1962) Meteorological techniques in shelter research. *In Symposium on Shelter Research, Aberystwyth,* 1962, pp. 45–66. Min. Agr. Fish & Food, London.

GLOYNE R.W. (1966) Instrumentation for field work on shelter research. Paper JSRC 9/66 prepared for Joint Shelter Research Committee. *Min. Agr. Fish & Food,* London. (limited distribution).

HILL L., GRIFFITH O.W. & FLACK M. (1916) Kata thermometers. *Phil. Trans. roy. Soc.* B. **207**, 186.

HOGG W.H. (1965) A shelter belt study—relative shelter, effective winds and maximum efficiency. *Agric. Meteorol.* **2**, 307–15.

HULL L.W.H. (1959) *History of the Philosophy of Science.* London

JENSEN M. (1954) *Shelter Effect.* Danish Tech. Press, Copenhagen.

LETTAU H.H. & DAVIDSON B. (1957) *Exploring the Atmosphere's First Mile.* Vol. I. *Instrumentation and Data Evaluation.* Pergamon Press, London.

LONG I.F. (1957) Instruments for micro-meteorology. *Quart. J. Roy. met. Soc.* **83**, 202–14.

MACCREADY P.B. (1964) Response characteristics and meteorological utilization of propeller and vane wind sensors. *J. appl. Meteorol.* **33**, (2), 182–93.

MARSHALL J.K. (1966) Automatic, battery-operated windspeed and direction recording system. Paper presented at UNESCO Symposium on Methods in Agroclimatology, Reading. (in the press).

METEOROLOGICAL OFFICE (1952) *Observer's Handbook.* H.M.S.O., London.

METEOROLOGICAL OFFICE (1956) *Handbook of Meteorological Instruments.* Part I. H.M.S.O., London.

MIDDLETON W.E.K. & SPILHAUS A.F. (1953) *Meteorological Instruments.* Univ. of Toronto Press.

NÄGELI W. (1953) Die Windbremsung durch einen grösseren Waldcomplex. *Proc. Congr. int. Union. For. Res. Organ., Rome.*

NÄGELI W. (1965) Über die Windverhältnisse im Bereich gestaffelter Windschutzstreifen. *Mitt. schweiz. Anst. forstl. VersWes.* **41**, 219–300.

OWER E. (1949) *The Measurement of Air Flow.* Chapman & Hall, London.

PLATT R.B. & GRIFFITHS J. (1964) *Environmental Measurement and Interpretation.* Reinhold, New York.

REIGNER I.C. (1964) How to make a wind-movement recorder from any spare drum-type recorder. *U.S. Forest Service Res. Note* USFS-NE-21.

RIDER N.E. (1960) On the performance of sensitive cup anemometers. *Met. Mag., London* **89**, 209–15.

ROGERS E.W.E. & WHITBREAD R.E. (1966) Industrial aerodynamics: a survey of research in the United Kingdom. *NPL Aero. Report* 1200.

ROSENBERG N.J. & ALLINGTON R.W. (1964) A microclimate sampling system for field plot and ecological research. *Ecology* **45**, 650–5.

RUTTER N. (1966) Shelter research at Univ. Coll. Wales, Aberystwyth. Paper JSRC 11/66, Joint Shelter Research Committee. *Min. Agr. Fish & Food, London.* (unpublished).

SCHMIDT R.L. & MARSHALL J.R. (1960) A wind-direction recorder for remote stations. *Ecology* **41** (3), 541–3.

SLATYER R.O. & MCILROY I.C. (1961) *Practical Microclimatology*. UNESCO.

SUMNER C.J. (1965) Long-period recorder for wind speed and direction. *Q. Jl R. met. Soc.* **91**, 364–7.

SUOMI V.E. (1957) Sonic anemometer. *In Exploring the Atmosphere's First Mile.* (Ed. H.H.Lettau & B.Davidson), pp. 256–66. Pergamon Press, Oxford.

TANNER C.B. (1963) *Basic Instrumentation and Measurements for Plant Environment and Micrometeorology*. Soils Bull. No. 6, Coll. Agric., Univ. Wisconsin, Madison, Wisconsin, U.S.A.

WOODRUFF N.P., READ R.A. & CHEPIL W.S. (1959) *Influence of a Field Windbreak on Summer Wind Movement and Air Temperature.* Tech. Bull. No. 100, Agric. Exp. Sta., Kansas State Univ., Manhattan, Kansas, U.S.A.

WORLD METEOROLOGICAL ORGANIZATION (1954) *Guide to International Meteorological Instrument and Observing Practice.* WMO-No. 8-TP.3, W.M.O., Geneva.

YAFFE C.D. Ed. (1956) *Encyclopaedia of Instrumentation for Industrial Hygiene.* pp. 627–31. Univ. Michigan Press.

APPLICATION OF INFRA-RED TECHNIQUES TO THE MEASUREMENT OF ENVIRONMENTAL FACTORS

S.D.SMITH, G.E.PECKHAM & P.J.ELLIS

J.J.Thomson Physical Laboratory, University of Reading

SUMMARY The limits for detection and wavelength resolution in the range 1–20 μ are used as a basis to discuss the future possibilities of infra-red methods for ecological problems. The practical limits are based on detectors suitable for field work, such as photoconductive cells and thermistor bolometers, and wavelength selection by interference filters and selective chopping. Instruments of such type can be small, rugged and relatively cheap. The performance is discussed specifically for such problems as the measurement of CO_2 in the presence of water vapour, determination of water vapour and other gases and the measurement of temperature of very small areas.

1 INTRODUCTION

Recent developments in infra-red techniques have indicated that small, portable and relatively cheap instruments for a variety of measurements such as CO_2, H_2O and other gaseous concentrations and both surface and air temperatures have become practical propositions requiring only fairly limited engineering development for realization.

This paper has been prompted by our experience in developing an infra-red satellite radiometer to sense atmospheric temperatures at different levels. The development of this technology has suggested several likely useful areas of application.

We commence by discussing some fundamental and practical limits for detection and wavelength selection. In this we discuss the wavelength range from 1–20 μ in which range can be found the bulk of the vibrational and vibration-rotational absorption lines of molecules, besides the peak of emission for temperatures of sources in the range 2000–200°K. We allow ourselves to be biased towards small rugged devices suitable for field use.

2 DETECTORS

Infra-red detectors are divisible into two classes: (i) thermal detectors, which are relatively slow, and (ii) quantum detectors which are fast. They are characterized by a factor of merit, called the detectivity ($D^\star$): this is defined from the expression

$$D^\star = \frac{1}{W}\frac{S}{N}\sqrt{(A\,\Delta f)} \qquad (1)$$

where S/N is the ratio of signal voltage to noise voltage when W watts are incident on a detector of area A and the amplifier bandwidth is Δf cycles/sec.

We shall discuss a small number of practical detectors; the performance limits are set by both fundamental and practical noise levels; in general for available uncooled detectors the fundamental limits are not reached, but some cases approach to within a factor of about 10. Any practical detector has a definite area, A, and for our discussion we find the *noise equivalent power* (N.E.P.) of a given detector (defined for signal = noise) a useful quantity. In equation 1 we therefore specify the bandwidth Δf as 1 cycle/sec.

The following then applies.

Thermal detectors

	Wavelength range	N.E.P. (W)	Area
(1) Golay Cell	All	$1{\cdot}8\times 10^{-10}$	3×3 mm
(2) Thermistor Bolometer	All	$1{\cdot}5\times 10^{-9}$	3×3 mm

Quantum: photoconductive

	Wavelength range	N.E.P. (W)	Area
(3) PbS	1–3 μ	3×10^{-12}	2×10 mm
(4) InSb	1–7 μ	$7{\cdot}0\times 10^{-9}$	$8\times 0{\cdot}5$ mm

There are many more detectors (see, for example Houghton & Smith (1966) Chap. V), but the above are the current practical devices we shall consider.

3 SOURCES

Tungsten filament lamps with glass (out to 3 μ) quartz (out to 4 μ) sapphire (to 6 μ) or periclase (MgO) (to 10 μ) windows may be used. Beyond 10 μ windows materials are more difficult and silicon carbide rods (Globars) or Nernst filaments are used, operating in air. A typical radiating area would be 1 cm$\times$2 mm, i.e. 0·2 cm^2. Operating at 2000°C we could have the following energies available in 1 cm^{-1} spectral band-width collected by an *f*/4 optical system, all of which could be condensed onto the detector.

5000 cm^{-1} (2 μ)	2000 cm^{-1} (5 μ)	667 cm^{-1} (15 μ)	wave number (wavelength)
3·2$\times$10^{-4}	9·3$\times$10^{-5}	3·3$\times$10^{-5}	watts

The spectral bandwidth has been chosen since the molecular absorption lines of such gases as H_2O, CO_2, CO, CH_4, NH_2, etc. are typically spaced by $\sim$1 cm^{-1}. It is apparent that given efficient spectral selection sufficient energy is available to make accurate measurements with a high degree of selectivity with respect to the molecular species if a bandwidth of a few cm^{-1} can be isolated.

Sources at room temperature or below may be of interest for emission measurements for the temperature of surfaces, restricted layers or regions of the atmosphere or again for concentration determination as follows:

Energy available in 1 cm^{-1} spectral bandwidth from 1 cm^2 area of source collected in f/4.

	10 μ	15 μ	wavelength
20°C	4·4$\times$10^{-7}	6·9$\times$10^{-7}	watts
0°C	3·1$\times$10^{-7}	5·5$\times$10^{-7}	watts
−73°C	4·3$\times$10^{-8}	1·4$\times$10^{-7}	watts

Again there is sufficient available energy for accurate measurement with the detectors of Section (2).

4 SPECTRAL SELECTION

Here we describe two techniques, essentially new in the method of their applications to this spectral region. These are interference filters and their combinations with selective choppers using gas cells. Both have the property of being able to pass large amounts of radiation by virtue of being able to accept large cone angles without loss of spectral performance (see Pidgeon & Smith 1964). The interference filters may be used alone; they consist of anything up to 40 layers of alternate high and low refractive index material in accurately controlled thicknesses (see Houghton & Smith (1966) Chap. VI). Design possibilities are considerable and three types occur—band pass of various widths, low pass and high pass. Band pass are most useful, for example, for gas analysis. A pass region of around 1 per cent of the centre frequency can be relatively easily attained, i.e.

5000 ($2\,\mu$)	2000 ($5\,\mu$)	667 ($15\,\mu$)	wave number (wavelenth)
50	20	6	bandwidth

Thus molecular line spacing is most easily approached at long wavelengths, where, unfortunately, detector, source and window problems are harder. This resolution is required in order to be *selective* in the gas examined. Higher resolution within the bandwidth of the interference filter can be obtained by selective chopping (and synchronous detection).

5 SOME INSTRUMENTS

We shall apply the building blocks so far described to the design of some simple but powerful instruments. We give three examples.

(a) *A filter gas analyser*

The conventional gas analyser as described by Luft is a wide band device and uses a rather large and insensitive detector. Its disadvantages are a lack of selectivity (e.g. CO_2 in presence of H_2O) and being a rather complicated apparatus of relatively high cost.

We have developed a very simple filter gas analyser using the narrow bandwidth made possible by the detectors and filters described in Sections 2 and 4. It has advantages in selectivity and simplicity and could be used for many gases.

The analyser is shown in Fig. 1. The source is a tungsten strip filament lamp with sapphire window. A 16 blade chopper which blocks exactly half the width of the beam makes an extremely simple double beam system without further optics. The beam then passes through a divided cell containing the gas to be analysed before being condensed onto an InSb detector via a filter centred on the $4{\cdot}2\,\mu$ absorption band. The CO_2

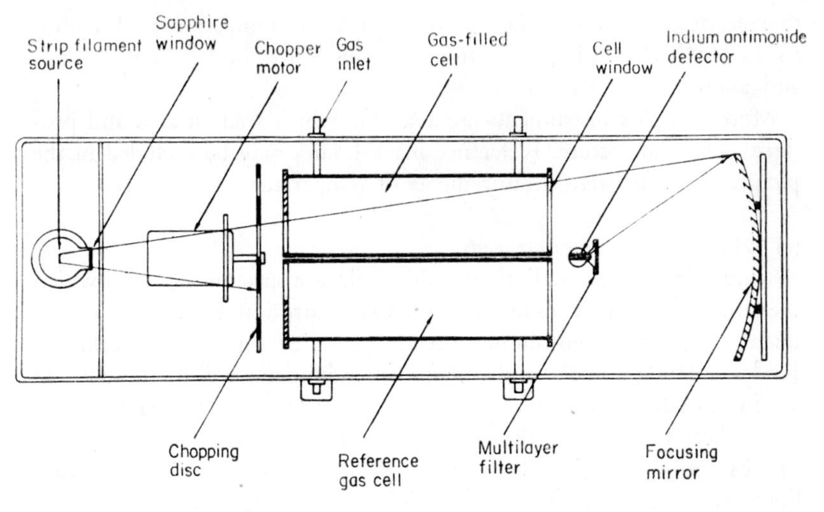

FIG. 1

absorption band extends from 2390 cm^{-1} to 2290 cm^{-1} and consists of about 90 lines. However the band is quite strong and well removed from any water vapour bands (at 2·6 μ and 6·5 μ). The device is thus quite independent of the presence of water vapour. Sensitivities down to 5 ppm have been obtained and can almost certainly be improved. The cost of the parts contained in the laboratory model can probably be kept below £100.

The device, which is described in British Patent No. 16011/64, is capable of development in several forms for continuous recording and can be readily adapted to other gases, e.g. H_2O, CO, NH_3 by using different filters.

(b) *Surface temperatures*

The radiant power from a surface can be measured easily using only a detector with, perhaps, some optical components to define the field of view and a chopper to modulate the radiation. Filters may be used to avoid regions of atmospheric absorption. Wide spectral bandwidths (e.g. 8–13 μ) may be used and there is little difficulty in collecting sufficient energy. It is difficult to interpret such measurements in terms of surface temperatures as some of the radiant power is reflected rather than emitted from the surface. However, temperature changes across a reasonably uniform surface can be accurately followed. The sensitivity of say a thermistor detector is sufficient to measure the temperature of 1 mm^2 to 2×10^{-4}°C (assuming 1 sec time constant, *f*/4 optical system, 8–13 μ, and approx 20°C source).

More complex instruments are available which scan an area and produce a thermal picture. Reference black bodies may be included in the picture if quantitative measurements of temperature are required.

(c) *Air temperature measurements*

We are working on a radiometer which selects a spectral region in which the atmosphere does absorb (15 μ CO_2 absorption band) and consequently, the instrument can measure atmospheric temperatures (Peckham, Rodgers, Houghton & Smith 1967). The radiometer will be mounted in a Nimbus satellite 600 miles above the earth's surface. The height in the atmosphere from which the received radiation was emitted depends on the absorption coefficient, being greater for larger absorptions. Weighting functions or the distributions in height of the origin of the radiation have been calculated for various spectral intervals in the neighbourhood of 15 μ. The spectral intervals must be of order 10 cm^{-1} so that the instrument can receive sufficient energy to make an accurate temperature measurement. Such an interval contains many spectral lines and consequently the absorption coefficient varies throughout the interval. This gives rise to a broadening of the weighting function.

The situation can be partly remedied by including an absorbing cell

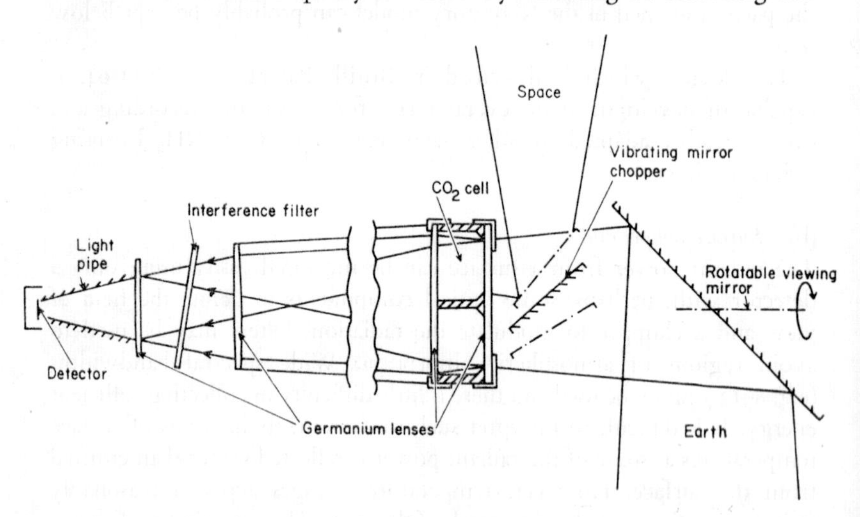

FIG. 2

of CO_2. This cell absorbs at the line centres and removes a high altitude tail that is otherwise present in the weighting functions.

The optical layout of the instrument is shown in Fig. 2. Radiation from the earth is modulated by a vibrating vane chopper. The chopper is in the form of a mirror directed toward space so that the instrument does not see radiation from the chopper itself. A lens system defines the field of view and contains the absorbing gas. The spectral interval is selected by a band pass filter and radiation is condensed onto a thermistor bolometer detector by a conical light pipe. A mirror can be turned so that the instrument sees space or a reference black body for calibration.

The instrument contains 6 channels measuring temperatures at different heights which are approximately evenly spaced up to 60 km. The accuracy should be better than 1°C.

The highest weighting functions are obtained by optically subtracting the signals in a double instrument, one half of which contains absorbing gas and the other half of which does not. By this technique, the instrument is made to respond only to those regions of the spectrum close to the line centres. An instrument of this type would have a range at ground level of only a few feet and might be used to record atmospheric temperatures in its immediate vicinity.

REFERENCES

HOUGHTON J.T. & SMITH S.D. (1966) *Infrared Physics*. Clarendon Press, Oxford.

PECKHAM G.E., RODGERS C.D., HOUGHTON J.T. & SMITH S.D. (1967) *Proceedings of the Symposium on Electromagnetic Sensing of the Earth from Satelites*. (Ed. Zirkind), Miami 1965. Polytechnic Press, New York.

PIDGEON C.R. & SMITH S.D. (1964) *J. Opt. Soc. Am.* **54**, 1459–66.

REFERENCES

WHAT FEATURES OF A MICROCLIMATE SHOULD BE MEASURED—AND HOW

J. W. Siddorn

Department of Zoology and Applied Entomology, Imperial College, London

Preliminaries

The distribution, behaviour and population dynamics of an organism in the field are all closely linked to the levels, ranges and cycles of climatic factors in its immediate environment. Which of these factors predominate will depend on the particular habitat at any time and on the state of development of the organism: a plant seed or an animal cyst will be less affected by variations and extremes of climate than will the actively growing stages.

It is therefore necessary in planning a field investigation of the ecology of an animal and/or plant community first to know enough of the life cycle(s) and the habitat(s) for an intelligent guess to be made as to the factors in the physical environment which are most likely to exert an appreciable effect. This information must be obtained either from previously published work on the species or on forms believed to be analogous, or from a preliminary field survey by the experimenter, or preferably from both.

Feasibility

It frequently happens that such an assessment indicates a range of factors greater than the range of detectors available for their measurement. One must then either invent an instrument (and a method for calibrating it!) or content oneself with the measurement of some dependent variable for which a method exists, or do without the measurement altogether. More often there are techniques in existence, but they may be too expensive, too time-consuming, too difficult to calibrate or maintain in calibration, too vulnerable to vandalism or attack by animals or extremes of weather, or dependent on unavailable mains electricity supply. Sometimes the apparatus manufacturer quotes a delivery date several months ahead with an (unquoted!) option to extend, at his will, for several more: this can be especially disastrous where the time available for the project is limited. Such short-term investigations are also vulnerable to catastrophic failure of elaborate recording equipment, and if too much reliance is placed on a multichannel, automatic, omniscient machine it may happen that the whole programme is stopped while the after-sales service department of the instrument manufacturer locates and corrects the fault. It pays to discuss the efficiency of apparatus suppliers in these respects with the users of their equipment before committing oneself too wholeheartedly to their products. Most are more reliable than some.

Automation

It is important to take a realistic view both of the time available for the purely physical aspects of the investigation and of the data handling facilities for their analysis. Contemplation of a vast range of possibly significant environmental stimuli sometimes leads to a despairing attempt to record everything for which an instrument is available, with the cosy feeling that multiple regression analysis in the long winter evenings will sort the wheat from the chaff. In the event this approach often leads to the accumulation of data more rapidly than they can be (or at least are) analysed, so that unexpected correlations between physical factors and biological variables are not noticed until it is too late in the season for confirmation. If the experimenter is in the unusual position of being able to command unlimited facilities in terms of apparatus and assistance, and if the data processing equipment is available when needed, it may happen that this omnivorous attitude to physical variables will shorten the period of an investigation. The great danger here is that the biological activities themselves might be less well observed than if more basic techniques are employed which entail the presence of the observer in the field.

At least in the early days of an ecological project, it may be better to concentrate on the biology and to restrict the physics to a combination of a few continuous records (or frequent observations in the absence of recorders) with indications of daily maxima, minima and totals, where appropriate simple equipment is available. Special care might be taken to include days where the climate is extreme as well as those on which it is 'average'. The biology should be kept in the forefront of thinking about the interpretation of the physical results: for example, the significance of minimum temperatures in a cold environment may be lessened if the biological process is subject to a developmental zero, or its analogues, above such a minimum.

Pitfalls

Surprisingly often, techniques are employed which themselves change the condition measured. One has seen wet-and-dry 'bulb' humidity devices operated in small volumes, and some thermistor thermometers are so designed that sufficient current passes through the thermistor to cause significant self-heating in air, whereas this source of error is negligible during calibration, usually performed with the element immersed in a liquid.

Almost too obvious to bear mention is that instruments should not casually be used for purposes for which they were not designed. For example, watch-form paper hygrometers of the Edney pattern are intended to be hygrometric memories rather than indicators of precise instant values of fluctuating humidities. The Edney hygrometer is normally calibrated at approximately the humidity it is desired to monitor and will then indicate departure from this humidity: its scale is merely for setting purposes and does not indicate the true humidity if this is far removed from that at which it was last calibrated. Yet many of these 'hygro-indicators' are used as hygrometers in varying humidities as thermometers are used to indicate varying temperatures, presumably because the dials are marked 'per cent R.H.'!

Large bulb thermometers used to measure air temperature in conditions of thermal radiation—sunshine or under laboratory lighting—are subject to a radiation error which results in an indicated temperature higher than ambient, the magnitude of the effect depending on the intensity of the radiation and on the size and albedo of the thermometer. Sometimes a tiny thermometer, perhaps a fine-wire thermocouple, may be the answer, but where the low thermal capacity of such a detector shows up the fine thermal structure of, for example, a small-scale turbulent wind, it may be impracticable to balance the indicator to make a reading. In this situation a radiation shield fitted to a thermometer of greater thermal capacity, and hence with a longer time constant, may be the best compromise.

Particularly insidious is the universal tendency to trust the readings of instruments, usually electrical, which give pointer-and-dial readings. One tends to record the indication of such a meter to an order of accuracy often unjustified by its capabilities, or at least its state of calibration. The best advice here is probably to develop an automatic distrust of dial readings unless and until a recent check calibration has been made and never to record the results to a greater standard of accuracy than is justified. Such routine tasks as battery standardization and thermocouple cold-junction checks must be scrupulously performed if undetectable and uncorrectable errors are not to creep into the results.

Light

The measurement of light intensity often gives rise to confusions. Light is a term for electromagnetic radiations which stimulate the human retina, and it shades off into infra-red in the long waves and ultra-violet in the short. Commercial light meters as used by lighting engineers have, for obvious reasons, a spectral sensitivity designed to match that of the human eye. The 'disappearing spot' type of photometer uses the human eye as a comparator even more directly. Now, unless the organism studied has the same spectral sensitivity as the instrument used, there must be an error in regressions of the meter readings on the biological effect. In many comparisons of light of constant quality, but varying intensity, the relative values will be usable, but if the spectral distribution of the emission varies, a photometer not matched to the biological process will give misleading results. For example, in studies involving insects, which commonly have a spectral sensitivity curve displaced towards the shorter wavelengths compared with the human eye, it would be injudicious to use an illumination meter to compare daylight levels in the field with artificial light in laboratory cages. There is no panacea: the instrument must be fitted to the circumstances, and full details given in published accounts of the research. Nowadays there is a considerable choice of quite different types of photoelectric detectors and these have widely differing spectral sensitivities. This makes the problem of matching, usually by the use of compensating filters, both more difficult and more necessary than when selenium barrier-layer cells, with the almost human spectral sensitivity (but for higher thresholds) were universally employed in light meters.

Apart from considerations of intensity and composition, light is also appreciated by an animal as a vector quantity. A dull day results in unorientated illumination compared with a sunny one or with a moonlit (or light-trapped!) night, and the importance of mere intensity is often secondary where compass orientation is involved. It may therefore be relevant to record aspects of light other than brightness and spectral quality.

Wind

Air movements present special difficulties on a small scale. Winds in the free atmosphere have a predominantly horizontal component, susceptible to measurement with cup and windmill types of anemometer and weather vanes. Near boundary layers, where organisms normally live, turbulence introduces large errors into the readings of such instruments, and the strength of the wind is reduced, often by an order of magnitude or more, to a level at which friction and inertia in the anemometer become a significant source of error. It is therefore necessary to determine, before choosing an anemometer, which wind-dependent functions in the organism are of importance in the investigation. For example, dispersion of flying insects may call for the use of such horizontal-component devices as those mentioned, whereas a sedentary organism may be affected more by ventilation or evaporation in which the direction of air movement is of secondary importance. In such a study, hotwire or thermistor anemometers may give more useful results.

Conclusions

The need for more sophisticated observations of microclimatic factors in ecological studies is becoming generally accepted. In many cases the unsolved problem is one of instrumentation, or the lack of it, and the challenge to devise small, inexpensive and trouble-free sensors is being met with increasing success. In the last analysis, however, what determines the value of the work is the intelligence with which the research has been planned and the results interpreted: an ecologist is more than the sum of his gadgets.

Source-books

Other papers in this Symposium cover the applied aspects of microclimatology and their bibliography. Of the many practical works on instrumentation and techniques the following may be found especially useful: Observers' Handbook, H.M.S.O. Publication M.O. 554; Handbook of Meteorological Instruments Part I, H.M.S.O. Publication M.O. 577; Environmental Measurement & Interpretation, Platt R.B. & Griffiths J., Chapman & Hall, London—and the International Journal of Biometeorology is rapidly becoming the repository of much valuable data on instrumentation and actual researches in climatology.

INTERCEPTION OF RAINFALL BY TREES AND MOORLAND VEGETATION

L. LEYTON, E. R. C. REYNOLDS & F. B. THOMPSON
Department of Forestry, Oxford University

SUMMARY The significance of interception of rainfall by plant covers is discussed from both biological and hydrological viewpoints. The measurement of interception loss, defined and usually determined as the difference between gross and net rainfall (throughfall plus stemflow), is subject to considerable instrumental and sampling errors and various methods for reducing these are described using as examples, the authors', studies in forest and moorland stands.

Interception loss is determined by the nature of the plant cover, the rainfall climate and the rate of evaporation of the intercepted water; these factors may be characterized respectively, by the canopy saturation value, the number, intensity and distribution of the rainstorms, and the duration of wetness of the canopy. Various methods are discussed for determining these quantities.

INTRODUCTION

In this paper we shall adopt the terminology of Hamilton & Rowe (1949), who define *interception* as the process in which rainfall is caught by the vegetative canopy and redistributed as throughfall, stemflow and evaporation from the vegetation. *Throughfall* is that portion of the rainfall which reaches the ground directly through gaps in the canopy and as drip from leaves and stems; that portion which reaches the ground by running down the stems is known as *stemflow*. *Interception loss* is defined as the portion of the rainfall retained by the above-ground parts of the vegetation and is either absorbed or evaporated into the atmosphere. *Gross rainfall* is the total rainfall incident on the canopy, whilst *net rainfall* refers to the quantity which actually reaches the ground and is therefore the sum of throughfall and stemflow.

Strictly speaking, in dealing with interception we should include, besides rainfall, other forms of precipitation, snow, fog, mist and dew, but as the measurement of these latter quantities and the interpretation of their ecological significance present many special problems beyond the immediate scope of this paper, we shall confine ourselves mainly to rainfall.

As a result of interception of rainfall the exposed surfaces of vegetation become more or less covered with a film or drops of water, and this is of importance from several aspects. For example, under British climatic conditions, the evaporation of intercepted water constitutes an appreciable fraction of the total loss of water by evaporation from a stand (Rutter 1963; Leyton, Reynolds & Thompson 1965; Reynolds 1966); since the quantity and persistence of this intercepted water may vary considerably from cover to cover, this may largely account for observed differences in the water consumption of these covers when soil moisture is not limiting. Many plants are able to absorb water directly into the leaves so that intercepted water may be of considerable survival value under conditions of soil moisture deficiency (Johnston 1964). The germination of, and subsequent invasion of the host by the spores of pathogenic organisms are often dependent on the presence and persistence of water films on leaf surfaces (Hirst 1957), whilst plant factors associated with interception are also of importance in the fate of growth regulators, pesticides etc. applied in the form of sprays. The effect of interception by a canopy on the amount and distribution of water reaching the ground below may also have far reaching effects on the nature and distribution of the ground vegetation, microbiological activity, root growth, soil differentiation and erosion. For example, with maize (Glover & Gwynne 1962) and with *Acacia* trees (Slatyer 1961), redistribution of intercepted water by stem flow may be of considerable importance for survival under arid conditions.

Thus, from both ecological and applied points of view, interception is a phenomenon of considerable biological and hydrological importance.

When dealing with the measurement of interception, two aspects have to be considered, namely the quantity of water retained on the plant surfaces (the interception loss) and the duration of wetness of these surfaces; the latter will be determined both by the interception loss and by various factors governing its rate of evaporation. Although occasional direct measurements have been made of interception loss, e.g. by weighing samples of vegetation before and after wetting (Rutter 1963) most measurements are made indirectly as the difference between gross and net rainfall.

The measurement of gross rainfall

For many purposes rainfall as collected by standard gauges is generally acceptable as a climatic index and data published by most national meteorological services based on networks of standard instruments (e.g. the 5 in. gauge of the British Meteorological Office) can be used. The problems of determining actual gross rainfall on an area basis are primarily ones of sampling, especially over areas in which large local differences occur. For interception and water balance studies where accurate estimates over areas are required a more intensive sampling is usually necessary than is normally covered by national networks and more attention has to be paid to reducing instrumental errors.

As far as sampling is concerned, area estimates can be improved either by increasing the number of standard gauges or by using gauges with larger collecting areas.

In hilly areas characterized by large local variations in rainfall, topographical stratification of gauges can be used in order to reduce the excessive numbers of gauges which may be required for adequate random sampling (Howe & Rodda 1960). But even with the best possible sampling system, the accuracy of rainfall measurements is influenced by other factors, especially instrumental errors, characteristic of most rain gauges (Reynolds 1963, 1964). By far the greatest source of error is due to the increase in wind speed and in turbulence about the gauge orifice, usually giving rise to decreasing catches and hence underestimates of the rainfall with increasing wind speed. This effect is modified by the general contour of the ground surrounding the gauge, the distance away and height of surrounding objects and by the height of the gauge above the ground. Great care must therefore be taken in siting the gauge (cf. Meteorological Office Handbook 1956). The effect of gauge height can be modified by means of turf walls or virtually eliminated by arranging the gauge orifice level with the surrounding ground surface, but with a suitably designed surround to prevent errors due to splash (Bleasdale 1958). Other errors may arise due to evaporation from the gauge itself in some climates, while in hilly terrain it has been shown both theoretically (Fourcade 1942) and in practice (Hamilton 1954) that it is incorrect to collect rain in a gauge with a horizontal orifice if the ground slopes appreciably.

In the case of wooded areas, gross rainfall may be sampled either in openings or by gauges mounted above the canopy. In the former case special attention must be paid to the size of the opening and the height of the surrounding trees. With above-the-canopy installations it is essential to reduce as far as possible the effect of increased wind speed and turbulence by fitting suitably designed shields around the gauges; in the absence of shields, gross rainfall may be seriously underestimated (Law 1957). Comparisons made by the authors over a number of years between standard 5 in. gauges fitted with different types of shield, unshielded gauges and a specially designed aerodynamic gauge, all mounted above the canopy, have confirmed the need for shields for consistent rainfall measurements in windy conditions (Reynolds & Leyton 1962, 1963; Reynolds 1964). Where wind speeds are low the catch of unshielded above-canopy gauges may be virtually identical with that of gauges in clearings (McCulloch 1962).

Theoretically, sampling by larger collecting areas should improve the accuracy of mean rainfall estimates. A particular example is provided by the authors' investigations using metre square collecting plots, i.e. plots cleared of vegetation and with the ground surface sealed by latex and surrounded by a splash zone, on exposed moorland in Yorkshire (Leyton, Reynolds & Thompson 1966). Comparisons with nearby standard 5 in. gauges for total catch over the period June to December 1966, showed close agreement between and within replicates; over shorter periods this agreement was not so good, the consistency between the standard gauges being rather better than that between the plots. Presumably because of greater errors involved in collecting and measuring the catch from the larger collecting areas, their theoretical advantages are not necessarily realized in practice. Perhaps the nearest approach to an absolute measurement of gross rainfall over a small area, which does not interfere with the vegetative cover, is the weighing lysimeter; however, because of the costs involved in constructing these and ensuring accurate measurements, their use must be restricted to special investigations.

Measurement of net rainfall

The problems of measuring net rainfall are to a large extent very different from those encountered with gross rainfall measurements; instrumental errors due to wind and evaporation are greatly reduced because of the generally more sheltered environment, but the sampling problem is increased considerably because of the large spatial variation in the rainfall commonly found below a vegetative cover.

If the area is small enough for practical purposes, but large enough to include a representative sample of the spatial variation, the whole of the net rainfall on the area can be collected and measured; in fact with grass and herb covers, it would be incorrect to introduce sampling since the size of most practicable collecting devices does not allow for strictly random sampling below the vegetation and hence is liable to give very biased results.

In the authors' investigations on moorland communities of heather, bracken and *Molinia*, the system used was to seal the surface of metre square quadrats with synthetic rubber latex and to measure the whole of the net rainfall as surface run-off (Leyton, Reynolds & Thompson 1966). As already implied in the case of similar rain collecting plots, the accurate measurement of surface run-off is by no means a simple matter, especially in remote areas where the system must be automatically recording and must also allow for considerable variations in the rate of flow. In the present system, 12 gallon storage drums are used fitted with floats, the positions of which are recorded every 3 min on magnetic tape (Limpet Logger); this allows for rainfall intensities of up to nearly 4 in./hr with a sensitivity of $\pm$0·002 in. As a check, the system has been designed to allow for visual readings of period totals.

With increasing plant size, i.e. with a shrubby or tree cover, the pattern of net rainfall variation is enlarged so that the collection of the total amount over an area large enough to be representative becomes much less practicable. Occasionally the whole of the net rainfall below single tree crowns has been collected and measured using for example, polythene sheet stretched below the crown (Wells 1963). Presumably this technique could be extended to cover a larger area of woodland; in practice however, it is much more convenient to sample by a network of gauges. These in addition often provide useful information on the distribution pattern of the net rainfall which is then sampled separately as stemflow and throughfall.

Measurement of stemflow

Many schemes have been described for collecting stemflow from trees, but in the authors' experience, a simple and effective technique is to wind a spiral of aluminium coach guttering around the stem, sealing the joints with bitumastic paint. In some species, stemflow is closely related to stem size; in such cases sampling can be improved, or the number of installations reduced, by taking this into account and stratifying. In other species, this relationship does not occur, and since stemflow can vary appreciably from tree to tree, the standard error of the mean value for the stand can be very large. In most cases however, the quantities of stemflow are small in relation to the throughfall, so that once this has been confirmed, stemflow can often be ignored; even when stemflow is included in the net rainfall, the standard error of the latter measurement is usually hardly affected. Nevertheless, from certain points of view, even a relatively small stemflow cannot be ignored since the quantities of water thus deposited around the base of the stem are, on an area basis, usually more than on any other region of the floor of the stand (Reynolds & Henderson 1967).

Relatively few measurements of stem flow have been made on plants other than trees. The authors have investigated the phenomenon in bracken both by attaching collecting cylinders to all stems on a metre square plot, and as the difference between net rainfall and throughfall. Under artificial rainfall of high intensity from a spray rig (up to 480 mm/hr for periods of 30 sec) and starting from the dry condition, stemflow in September ranged from 17·3 per cent of a net rainfall of 1·15 mm, to 19·1 per cent of a net rainfall of 2·46 mm; when the bracken foliage was already nearly saturated, the absolute amount of stemflow was higher, but its proportion of the net rainfall decreased. This latter result is not unexpected since with the 'interception storage capacity' (defined below) of the canopy almost satisfied, the interception loss decreases and throughfall increases more than stemflow.

In November, when the bracken was dead but still standing, stemflow increased to 31 per cent of a net rainfall of 1·05 to 1·19 mm, again using artificial rain. Though high this is by no means one of the highest values recorded for stemflow in plant stands; Beard (1962) found a stemflow of 51 per cent of net rainfall in grass, 2·5 ft high. However, its importance as a pathway for water reaching the ground must decrease with increasing number of stems per unit area. With bracken, under the rainfalls mentioned above, stemflow gave a concentration of water over the 1 cm^2 around the stem base some 37–82 times the average throughfall for a mean density of 1 stem per 250 cm^2 of ground surface. If similar stemflow and throughfall values are applied to heather covers with an average of 1 stem per 31 cm^2 ground surface, the concentration of water around the stem bases would be only seven times that of the average throughfall.

Measurement of throughfall

The problem of sampling throughfall in forest stands has been discussed in some detail by Reynolds & Leyton (1963). Under most plant covers the spatial variation in throughfall, as determined by vegetative characteristics, is of three kinds: (1) a systematic variation below the crowns of individual plants or clumps of plants constituting the stand; under certain trees for example (cf. Leyton & Carlisle 1959), there is a practically linear increase in throughfall from the stem to below the edge of the crown, where there may be a sharp increase due to drip; (2) quantitative differences between the basic patterns below individual plants due to differences in crown size, height, shelter etc. and (3) gaps in the canopy through which rainfall penetrates directly to the ground. All three sources of variation, when present, must obviously be taken into account when sampling for throughfall, but their extent varies considerably from stand to stand. For example, in one case of a broad leaved woodland with closed canopy, the spatial variability was so small that only 4–5 randomly distributed standard gauges were found necessary to give a reliable estimate of mean throughfall; at the other extreme, for the same area of spruce plantation, also with a closed canopy, the spatial variability was so great that even with 20 standard gauges, distributed at random, the standard error of the mean was 14 per cent. Apart from still further increasing the number of gauges the accuracy of throughfall estimates based on sampling can be increased in various ways: (1) by determining the nature of the systematic variation of the throughfall in relation to the location of the gauges below the crown and applying the resulting regression to the evaluation of mean throughfall (Reynolds & Leyton 1963); (2) by increasing the number of sampling points by moving the gauges to new positions, e.g. after certain quantities of rain (Wilm 1943); this has the disadvantage that the mean throughfall can only be calculated after a number of new positions has been used so that it cannot be applied to short term estimates; and (3) the most practicable, by increasing the size of the collecting areas to integrate more completely the pattern of throughfall; as in the case of the rain collecting plots mentioned earlier, this method introduces the problem of accurately measuring often very large quantities of water. Using approximately 60×90 cm troughs, the authors have attained some success using tipping buckets operating electro-magnetic counters which can be read at a glance. Large collecting devices however, have a more serious disadvantage in that their size prohibits a truly random distribution and they might therefore give biased results, e.g. the region close to the stems cannot be adequately sampled. In one of our forest studies (Leyton, Reynolds & Thompson 1965) this limitation has been overcome by sampling this region separately using annular gauges mounted around random tree trunks. The shape of the large gauges is of some importance in sampling. Long narrow troughs (Eidmann 1959) would be the best for integrating the pattern of throughfall variation, but are the most difficult to randomize; large circular gauges are the easiest to randomize but would only be more efficient for sampling than small gauges in the case of purely random variations.

Interception loss

As defined before, interception loss is the difference between gross and net rainfall and has frequently been expressed simply as a percentage of the gross rainfall on an annual or seasonal basis. This however, is of limited value since it gives no indication of the effect of the distribution in time of the rainfall which can significantly alter the interception loss of identical plant covers. A better way is to express the interception loss as a regression on gross rainfall in individual storms (Aranda & Coutts 1963; Rowe & Hendrix 1951), but this still does not allow for extrapolation to either different canopy densities of the same species, or different rainfall climates. For this, some relevant parameter of interception must be chosen, characteristic only of the plant cover and excluding the interaction between the cover and the climate. The most appropriate parameter is the *interception storage capacity* which may be defined as the equivalent depth of rain held by the aerial parts of the vegetation when it is fully wetted. Apart from losses due to evaporation and absorption during rain, this must represent the maximum amount of interception that can occur during any single rain storm and hence must be the dominant vegetative parameter determining interception characteristics. We have discussed elsewhere methods for determining this quantity both by artificial wetting and under natural rainfall conditions (Leyton, Reynolds & Thompson 1965). Once the interception storage capacity has been determined for a particular species, it can be estimated for other stands of the species, e.g. with different cover densities, on the basis of relative leaf area indices, or similar expressions of the amount of plant cover, e.g. by the 'equivalent film thickness' method of Merriam (1961). The interception storage capacity combined with data on local rainfall climates can then be used for estimating likely interception losses of different plant covers in different areas.

Duration of wetness

In relating interception loss to climate, a knowledge of the duration of wetness of the plant cover is necessary since it determines how much of the interception storage capacity will be available at the beginning of a particular rain storm. This is particularly important during showery periods with little evaporation of intercepted water during the intervals between successive showers. A number of instruments for recording the duration of leaf wetness are described by Schnelle, Smith & Wallin (1962). Most of the instruments in current use were designed primarily for plant pathological studies and with few exceptions record the wetness of artificial surfaces either mechanically (through weight or dimensional changes) or electrically (through changes in resistance with wetting). If artificial surfaces are used there is the problem of ensuring that their behaviour in wetting and drying is identical with that of the natural plant surfaces being studied. None of the standard designs really satisfy these requirements except possibly within a very narrow range of application which has to be confirmed empirically by direct comparison with the plant surface concerned.

Wells (1963) found fairly close agreement in the laboratory between the wetting and drying of spruce foliage and changes in weight of an expanded polystyrene block which forms the basis of the Hirst (1957) wetness recorder; a number of these instruments was then used to follow wetting and drying regimes at different points within a 35 ft high Norway spruce stand near Oxford. They were found to be sufficiently sensitive to record dewfall as well as the effect of rain. Some of the results for rainfall wetting and subsequent drying have been published elsewhere (Leyton, Reynolds & Thompson 1965); the results for dewfall and its evaporation, hitherto unpublished, are of some interest in showing the value of measurements of this kind. Over the period March to July 1962 (inclusive), one instrument at 25 ft between the tree crowns recorded dew on 42 per cent of the days with an everage accumulation period of 5·4 hr; maximum deposits were obtained about 2 hr after sunrise with a mean value of 0·5 g and took on average 3·1 hr to evaporate. At a height of 35 ft, level with the top of the trees, about twice as much dew was collected in the same period, but as expected, this evaporated much more rapidly; at 18 ft within the crown, no dewfall was ever recorded. Although this instrument appeared to give consistent results, it had a number of drawbacks; apart from its large size, it was insufficiently damped for windy locations and its range was too limited in relation to the size of the polystyrene block to record both complete wetting and drying of the block.

An alternative approach to the problem is to record changes in wetness on the actual plant surface itself. Certain techniques have been described based on changes in resistance between electrodes applied to leaf surfaces,

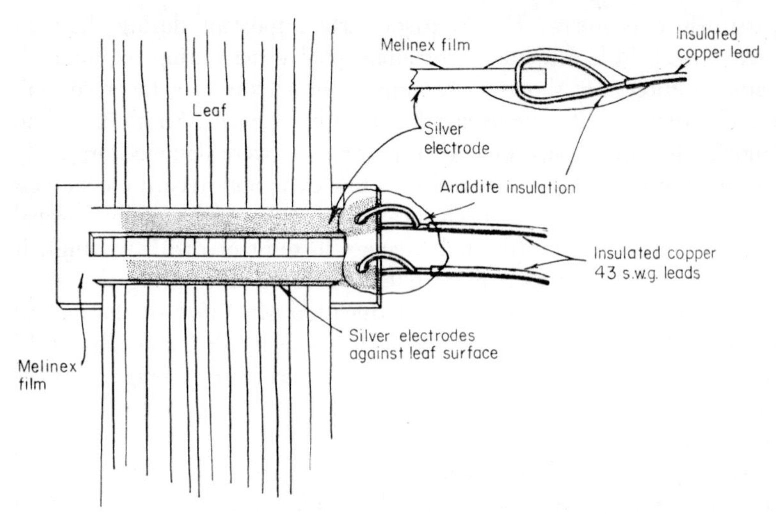

FIG. 1. Leaf wetness unit.

but may have important drawbacks, e.g. that of Woolford & Smith (1963) has to be frequently moved because of physical damage to the tissues. One of the present authors has improved the design by painting silver electrodes on a thin plastic film which is fastened against the leaf (Fig. 1). Though this is not ideal for all leaf shapes it does avoid damage, is easily attached and has little effect on the microclimate of the leaf, an important requirement with these techniques; electrodes painted directly on to the leaf surface would be even better and are currently being tested. Using an alternating current bridge circuit and amplifier, data on changes of resistance due to wetting can be recorded on magnetic tape along with those on relevant microclimatic factors; alternatively an 'event recorder' can be incorporated so as to identify the time at which the surface reaches a predetermined level of wetting.

Ideally the most suitable technique would be one which records the degree of wetting without any attachment to the plant. The authors have made some tentative, but so far inconclusive, studies along these lines using changes in the albedo of the surfaces in natural daylight; possibly a more sensitive method would be to use monochromatic light.

Conclusions

The process of interception is far more complex and its significance far wider reaching than many workers in the past appreciated. Because of faulty measurement techniques, both instrumental and sampling, much of the data published in the literature is of doubtful value; furthermore, the data are often expressed in such a way that the conclusions at best, are only of purely local interest. In this paper we have attempted to restrict the treatment of the subject to essential features and have omitted details of particular studies except where they have been thought necessary to illustrate techniques. In this way we hope to have emphasized sufficiently the importance of methodology and the need to present the data with appropriate statistics of their reliability.

REFERENCES

ARANDA J.M. & COUTTS J.R.H. (1963) Micrometeorological observations in an afforested area in Aberdeenshire: rainfall characteristics. *J. Soil Sci.* **14**, 124–34.

BEARD J.S. (1962) Rainfall interception by Grass. *Jl S. Afr. For. Ass.* **42**, 12–15.

BLEASDALE A. (1958) A compound raingauge for assessing some possible errors in point rainfall measurements. *Hydrol. Memor. No.* 3.

EIDMANN F.E. (1959) Die Interception in Büchen-und Fichtenbeständen; Ergebnis mehrjähriger Untersuchungen im Rothaargebirge (Sauerland). *U.G.G.I., Ass. Int. Hydrol. Sci., Symp. Hannoversch. Munden* Publ. No. 48, 5–25.

FOURCADE H.G. (1942) Some notes on the effects of the incidence of rain on the distribution of rainfall over the surface of unlevel ground. *Trans. R. Soc. S. Afr.* **29**, 235–54.

GLOVER J. & GWYNNE M.D. (1962) Light rainfall and plant survival in East Africa. 1. Maize. *J. Ecol.* **50**, 111–8.

HAMILTON E.L. (1954) Rainfall sampling on rugged terrain. *Tech. Bull. U.S. Dep. Agric.* No. 1096.

HAMILTON E.L. & ROWE P.B. (1949) Rainfall interception by chaparral in California. *Calif. For. Range Exp. Sta.* 46 pp.

HIRST J.M. (1957) A simplified surface-wetness recorder. *Pl. Path.* **6**, 57–61.

HOWE M.G. & RODDA J.C. (1960) An investigation of the hydrological cycle in the catchment area of the River Ystwyth during 1958. *Wat. & Wat. Engng.* **64**, 10–16.

JOHNSTON R.D. (1964) Water relations of *Pinus radiata* under plantation conditions. *Aust. J. Bot.* **12**, 111–24.

LAW F. (1957) Measurement of rainfall, interception and evaporation losses in a plantation of Sitka spruce. *U.G.G.I., Ass. Int. Hydrol. Sci., Ass. Gen. Toronto,* 1957, 2, Publ. No. 44 (1958).

LEYTON L. & CARLISLE A. (1959) Afforestation and water supplies in Britain. *Mitt. schweiz. Anst. forstl. VersWes.* **35**, 51–4.

LEYTON L., REYNOLDS E.R.C. & THOMPSON F.B. (1965) Rainfall interception in forest and moorland. *In Forest Hydrology*, Edit. Sopper W.E. & Lull H.W. Pergamon, Oxford, 163–77.

LEYTON L., REYNOLDS E.R.C. & THOMPSON F.B. (1966) Hydrological relations of forest stands. *Report on Forest Research* 1965, 119–23. H.M.S.O. London.

McCulloch J.S.G. (1962) Development of softwood plantations in bamboo forest. The hydrological analysis measurements of rainfall and evaporation. *E. Afr. agric. For. J.* **27** (Special Issue), 88–92.

Merriam R.A. (1961) Surface water storage on annual rye grass. *J. geophys. Res.* **66,** 1837–8.

Meteorological Office (1956) *Meteorological Observer's Handbook.* 2nd edit. H.M.S.O. London.

Reynolds E.R.C. (1963) Comparisons of rain-gauge measurements. *Met. Mag., Lond.* **92,** 210–3.

Reynolds E.R.C. (1964) The accuracy of rain-gauges. *Met. Mag., Lond.* **93,** 65–70.

Reynolds E.R.C. (1966) The hydrological cycle as affected by vegetation differences. *Advmt. Sci., Br. Ass.,* Section K*, Nottingham, 1966.

Reynolds E.R.C. & Henderson C.S. (1967) Rainfall interception by beech, larch and Norway spruce. *Forestry* **40,** 165–84.

Reynolds E.R.C. & Leyton L. (1962) The measurement of rainfall in woodland. *Wat. Res. Association.* Special Report. S.R. 2 (4), 1–8.

Reynolds E.R.C. & Leyton L. (1963) Measurement and significance of throughfall in forest stands. *In Water relations of Plants,* Ed. Rutter A.J. & Whitehead F.H. Blackwells, Oxford, 127–41.

Rowe P.B. & Hendrix T.M. (1951) Interception of rain and snow by second-growth Ponderosa Pine. *Trans. Am. geophys. Un.* **32,** 903–8.

Rutter A.J. (1963) Studies in the water relations of *Pinus sylvestris* in plantation conditions. I. Measurements of rainfall and interception. *J. Ecol.* **51,** 191–203.

Schnelle F., Smith L.D. & Wallin J.R. (1962) Report on instruments recording the leaf wetness period. *Agrarmet. Komm. WMO.* Doc. 34, Toronto.

Slatyer R.O. (1961) Methodology of a water balance study conducted on a desert woodland (*Acacia aneura* F.Muell.) community in Central Australia. *Plant-water relationships in arid and semi-arid conditions. Proceedings of the Madrid Symposium.* Paris. U.N.E.S.C.O. 15–26.

Wells K.F. (1963) *Some Aspects of the Role of the Forest Canopy in the Fate of Rainfall.* B.Sc., Thesis, Univ. Oxford.

Wilm H.G. (1943) Determining net rainfall under a conifer forest. *J. agric. Res.* **67,** 501–12.

Woolford M.W. & Smith J.D. (1963) Recording wetness on rye grass (*Lolium*) leaves. *N.Z. Jl agric. Res.* **6,** 578–84.

MEASUREMENT OF RADIANT ENERGY

G. SZEICZ
Rothamsted Experimental Station, Harpenden, Herts.

SUMMARY Total solar radiation T falling on a horizontal surface can be divided into two components: S direct radiation from the sun and D diffuse radiation from the sky. A different, but for ecologists more important division, is by spectral regions: the solar spectrum in the 0·3–3·0 μ band can be split into visible or photosynthetically active radiation between 0·4–0·7 μ and near infra-red radiation between 0·7 and 3·0 μ.

Atmospheric water vapour, carbon dioxide and the surface of the earth also exchange radiation with an intensity proportional to the fourth power of their absolute temperatures. This radiation is commonly referred to as long-wave exchange; the spectrum extends from about 3 μ to 80–100 μ.

Most of the net radiant energy, the balance between all the incoming and outgoing components, when absorbed by plants and animals is transformed into heat. To dispose of this large amount of energy—about 0·5 kw/m^2 in bright sunshine—plants transpire, animals sweat and both expose themselves to the cooling wind. Between 0·4 and 0·7 μ some of the absorbed radiation is used in photosynthesis and a small fraction of it is stored in the end products. The human eye is also sensitive to this region but with a different and well defined spectral response.

Incoming and reflected solar radiation are usually measured by thermopiles whose surfaces are protected from wind and weather by glass domes. Visible radiation can be measured by combining the thermopiles with cut-off glass filters. Net radiation can be measured with blackened flux-plates which are either protected by polythene domes or ventilated by a fast jet of air that swamps the effects of the wind. Inside crops where radiation is uneven, long tubular solarimeters and net radiometers can be used to give good spatial averages.

Solarimeters are calibrated against sub-standards in Meteorological Office networks and net radiometers in turn can be calibrated against solarimeters by 'shading'. When photocells are calibrated against solarimeters in daylight, caution is necessary when using them in artificial light. Because of their uniform spectral response thermopile instruments read in absolute energy units but photocells in units of luminous energy. The two are comparable only when the spectral composition does not change appreciably with intensity.

To obtain consistency and good quality in measurements it is often worthwhile to seek advice from meteorologists or physicists on the availability of measurements in the national network system and on facilities for calibration.

INTRODUCTION

Solar radiation reaching the surface of the earth is the source of energy for all kinds of life. Plants use it when they assimilate carbon dioxide, and animals absorb it to keep themselves warm. However, while plants are absorbing radiation to keep photosynthesis going, they also absorb a lot of surplus energy that in turn is transformed into heat. They keep their leaves cool by exposing them to the wind and evaporating water through their stomata. In their everyday activities, animals too may be exposed to excessive radiation loads. When they get too hot, the smaller animals, if they can, simply seek out a comfortable microenvironment. The larger ones have to sweat and cool themselves by the wind, but mostly re-radiate the surplus heat.

The energy of solar radiation usually far exceeds what life on earth needs. Because it is such a large and important parameter in ecology, it needs to be measured accurately and reliably.

RADIATION COMPONENTS

The sun emits radiation as a black body at around 6000°K; as a result the solar spectrum extends from about 0·3–3·0 μ, with a sharp peak at 0·55 μ. In contrast, the earth as a planet radiates in the long-wave region like a black-body at about 250°K and its spectrum extends from about 3–100 μ (Fig. 1). Hence the two kinds of radiation, solar and terrestrial, can conveniently be separated.

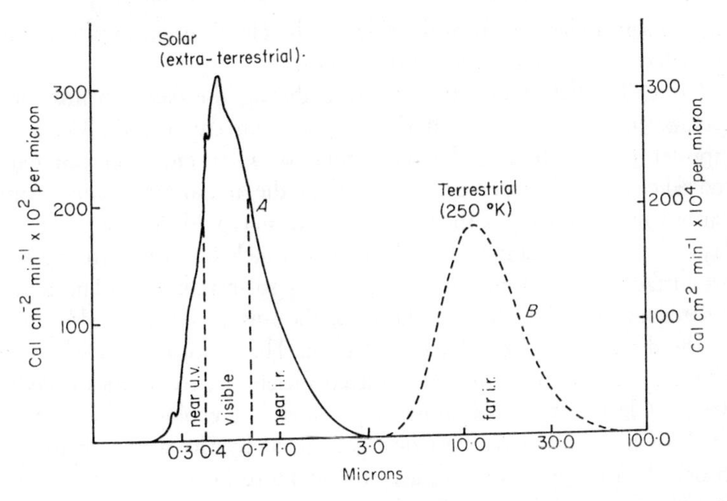

FIG. 1. Spectral distribution of extraterrestrial solar radiation (*A*) and of black-body radiation at 250°K (*B*), based on data from Smithsonian Meteorological Tables (1951). Note that wavelength is on a logarithmic scale and units for *A* are 100 times larger than those for *B*. (From Blackwell 1966.).

Solar radiation (0·3–3·0 μ)

At the earth's surface, solar radiation includes a direct component, intensity S, coming from the direction of the sun, and a diffuse component, intensity D scattered downwards by air and gas molecules, dust particles, water droplets, or reflected from the sides of the clouds, coming from the whole hemisphere. The total, $T = S + D$, is often referred to as solar radiation, short-wave radiation, global radiation, or very loosely, as sunshine. Three important questions concern the ecologist. First, how much energy is delivered on unit area in unit time? (Strictly this should be referred to as the 'irradiance' but the term intensity is commonly used.) Secondly, how does this intensity change with time of day and season? Thirdly, how is the energy distributed in the solar spectrum? To give orders of magnitude, the solar constant outside the atmosphere, measured perpendicular to the sun's rays, is 2·0 cal cm^{-2} min^{-1}. In the tropics with the sun overhead, atmospheric attenuation decreases this intensity at sea level on a horizontal surface to about 1·6 cal cm^{-2} min^{-1}. In Britain, in midsummer, with the sun about 60° above the horizon in a clear sky, the maximum intensity occasionally reaches 1·3 cal cm^{-2} min^{-1}. In the presence of a moderate amount of white cloud, not obscuring the sun, reflection from the sides of the clouds often increases this to 1·5–1·6 cal cm^{-2} min^{-1} for short periods.

The daily *total* of radiation received during the day depends on the average intensity throughout the day, and on day length. On a clear summer day in Britain the daily total of solar radiation can exceed 700 cal cm^{-2}, a few per cent more than the maximum many tropical stations record during a shorter day. In winter, with low sun and short days, the British maximum daily total is much less, only 100 cal cm^{-2}. All surfaces reflect some of the incoming solar radiation. The amount is determined mainly by the nature of the surface but may also depend on the direction and quality of radiation. The ratio of reflected to total incoming radiation is the reflection coefficient, α, sometimes called the albedo. Clearly, α is an important parameter in ecology, showing how much of the short-wave radiation is rejected by the surface before it is absorbed. Although most surfaces reflect more in some spectral regions than in others—giving a surface its colour—in terms of energy, reflection coefficients usually refer to the ratio of reflected to total radiation in the whole 0·3–3·0 μ band.

Terrestrial radiation (3–100 μ)

Most natural objects on the surface behave almost like black-bodies, and radiate energy in proportion to the fourth power of their absolute temperatures. Atmospheric gases radiate in discrete wavebands and, in addition, long-wave radiation will be emitted by water droplets and ice crystals in clouds.

Thus at the surface of the earth there are two components in a long-wave radiation exchange: a downward component, emitted by the atmosphere, depending on air temperature and vapour pressure, and an upward component, emitted by the ground, depending on surface temperature. To give orders of magnitude, the average daily total of downward atmospheric radiation over Europe, taking 276°K (3°C) as the mean temperature of the lower layers of the atmosphere, is $L_d = 680$ cal cm^{-2}. Upward emission from the ground is always larger, and taking a mean temperature of 286°K (13°C), $L_u = 780$ cal cm^{-2}. Although these are large quantities the *net* long-wave loss from the surface, $(L_u - L_d)$, is only 100 cal cm^{-2} day^{-1}, or 0·070 cal cm^{-2} min^{-1}. On an overcast day with low clouds and little surface heating, the cloud base and the ground are almost at the same temperature, so that the net loss is zero.

Net radiation (0·3–100 μ)

Net radiation is the difference between the short and long-wave radiation gained by the surface and that lost by reflection and long-wave emission.

$$R = T - \alpha T - L_u - L_d$$

In radiation climatology this is the most important term, because net radiation is the energy available for evaporation, melting snow and heating the soil and the air.

RADIATION IN ECOLOGY

In ecology, radiation is important in two different ways. First, independently of wavelength, all radiation can be transformed into heat. Secondly, radiation affects photosynthesis in the 0.4–0.7 μ range. When a leaf absorbs carbon dioxide, radiation absorbed in this waveband supplies the energy for photosynthesis. Thus whereas transpiration of crops and the heat-balance of animals is determined by the net exchange of radiation, photosynthesis depends on the receipt of radiation in the visible region.

In plant ecology, radiation is often measured with photometers, but this is misleading because the spectral sensitivites of the human eye and the photosynthetic mechanism are very different. Suppose in a growth-room with fluorescent tubes the intensity of illumination in photometric units is 3000 ft candles. If the tubes had a ratio of *visible* radiation to illumination (calculated from the spectral curves) of $4{\cdot}75 \times 10^{-5}$ cal cm^{-2} min^{-1}/ft candle (e.g. Phillips TL 33/40 W), then the energy output would be 0·142 cal cm^{-2} min^{-1} of visible radiation, approximately 25 per cent of full midsummer sunshine. However, because of the different spectral composition, sunshine with the visible radiation intensity of 0·142 cal cm^{-2} min^{-1} would register as 2300 ft candles on a photometer. Table 1 (Gaastra 1966*a*) shows the variation of photosynthetically useful energy content per unit luminous flux for different light sources.

The spectral sensitivity of the human eye is very closely represented by the standard curve of vision to which the response of photometers can be accurately matched (Fig. 2, p. 117). Photometers are easy to use, but caution is necessary when interpreting their readings for other animals whose spectral sensitivity of vision is not known, as in studies of the response of insects to very weak light, such as moon-light or star-light. Obviously, it is very difficult to measure and define a threshold value.

TABLE 1. Relation between energy and photometric units for various lamps. The factor 'f' is the energy in the waveband 0·4–0·7 μ expressed in cal cm^{-2} min^{-1}/1000 ft candles. TL; Phillips fluorescent tubes; 29, warm white; 32, warm white de luxe; 33, white; 34, white de luxe; 55, daylight; 57, daylight special. HPL; Phillips high pressure mercury vapour lamp, with fluorescent coating. Xenon; Osram Xenon lamp. (Goastra 1966a).

Lamp type	f
TL-29/40 W	0·0435
TL-32/40W	0·0571
TL-33/40 W	0·0475
TL-34/40 W	0·0573
TL-55/40 W	0·0573
TL-57/40 W	0·0650
HPL/400 W	0·0465
Xenon/600 W	0·0655
Sun	0·0618

Different sensing devices can be tried but perhaps the most promising ones are the new photoresistors (lead sulphide and cadmium sulphide etc.) Very recent work in the USA, however, showed that at night some insects receive and transmit information not by visible light but in a very narrow wave band of the far infra-red, acting like sensitive bolometers (Callahan 1965).

UNITS

The fundamental unit of energy income on a unit area is erg cm^{-2} sec^{-1}, but radiation measurements are often expressed in a larger unit in mW cm^{-2}. However, in heat and water balance studies a more familiar unit, the calorie is widely used. This is the heat equivalent of energy. The conversions are:

$$1 \text{ mW cm}^{-2} \equiv 10^4 \text{ erg cm}^{-2} \text{ sec}^{-1}$$
$$1 \text{ cal cm}^{-2} \text{ min}^{-1} \equiv 69{\cdot}8 \text{ mW cm}^{-2}$$

Sometimes radiation is expressed as the water equivalent of the latent heat of evaporation, in mm of water. This depends slightly on temperature: at 20°C it is 585 cal g^{-1} or 58·5 cal cm^{-2} per mm of water. Hence:

$$1 \text{ cal cm}^{-2} \equiv 1{\cdot}71 \times 10^{-2} \text{ mm of evaporation}$$
$$1 \text{ cal cm}^{-2} \text{ min}^{-1} \equiv 1{\cdot}03 \text{ mm h}^{-1}$$

Photometers are calibrated in units of illumination; e.g. either the metric unit: lumens m^{-2} (lux), or the British unit: lumens ft^{-2} (ft candles). The relation is: 1000 lux (1 kilolux) = 93 ft candles. When photometers are used outdoors, Table 2 gives approximate relationships for different seasons relating illumination to *total* solar radiation measured on a horizontal surface (calculated from Blackwell 1966).

TABLE 2. Luminous efficiency of solar radiation at Kew (in ft candles/cal cm^{-2} min^{-1})

Winter	Summer	Mean	State of sky
7400	8400	8100	cloudless
6200	7800	7200	overcast
6900	8100	7500	average conditions

TYPES OF INSTRUMENTS AVAILABLE

Radiation measurements in ecology can be divided in the following way:

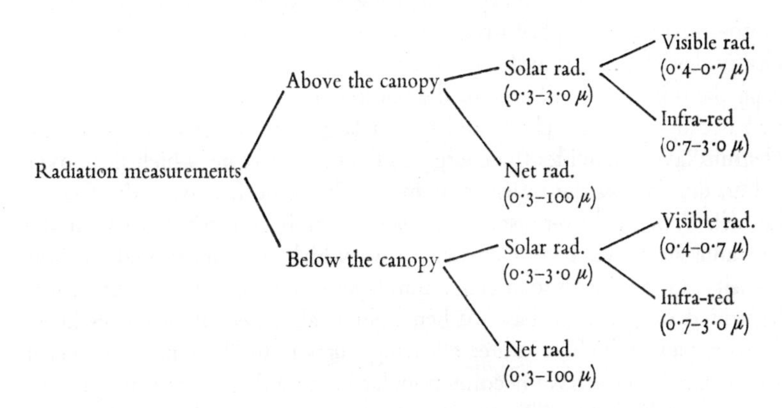

The two major groups of instruments designed for work above and below the canopy differ in size and geometry. Above the canopy, where the field of radiation is uniform, a small sensing element is an asset because it can be light and portable. However, radiation in the canopy is very uneven; sunflecks move as the plants move and the sun changes position.

Thus it is important to sample a large area at a given height to get a good spatial average and so tube solarimeters and strip net radiometers have been developed.

Radiation sensors can employ one of the following principles:

(a) thermoelectric
(b) photoelectric
(c) bimetallic distortion
(d) distillation
(e) photochemical

(*a*) *Thermoelectric* radiation instruments employ thermopiles, which are essentially a group of thermocouples arranged in series so that the two sets of opposing junctions are at different temperatures. One set, the hot junctions, is blackened and exposed to the radiation; the opposing set can either be arranged in good thermal contact with the body of the instrument (as in the Kipp solarimeter) or grouped under a white or reflecting cooler area (as in the Eppley and Rothamsted solarimeters). With a thermally well-designed instrument the resulting temperature difference, and hence the emf output, is a linear function of the radiation intensity, and its blackened sensor is uniformly sensitive to all wavelengths (Monteith 1959) (Fig. 2a).

(*b*) There are three main types of *photoelectric* devices, the photo-voltaic, photo-emissive, and photo-resistive cells. They have different spectral sensitivities (Fig. 2b, d and e respectively) and caution is necessary when using them in biology. Some of the photo-voltaic cells, e.g. selenium, are used because they are cheap and give a large output. Here light on the selenium layers provides the energy to liberate electrons which flow over and accumulate in the transparent metal film deposited over the face of the cell. When a low-resistance galvanometer is connected between the film and the metal base a current flows which is proportional to light intensity. In the silicon solar cell, a thin layer of *n* type of semiconductor is desposited on a *p* type base. When light is absorbed, it produces hole-electron pairs which separate, allowing current to flow in an external circuit. Lately these have become popular because their output per surface area is large (Federer & Tanner 1965).

In photo-emissive cells or photo-tubes the electrons are liberated from the sensitive surface of the cathode into the surrounding gas or vacuum and are collected on the anode. With a steady applied voltage the current is proportional to light intensity.

In photo-resistors the electrical resistance of the cell changes with light

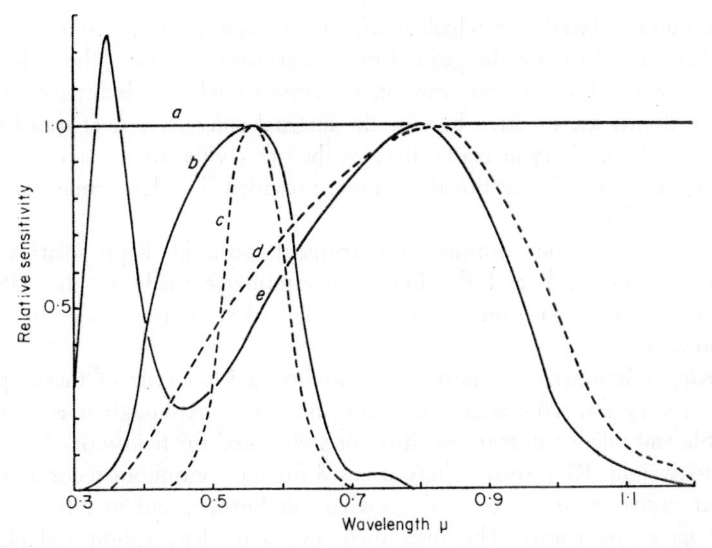

FIG. 2. Spectral sensitivity of radiation meters: *a*, thermopile, *b*, selenium cell; *c*, selenium cell, corrected to match the sensitivity of the human eye; *d*, silicon solar cell; *e*, phototube with Ce–Sb cathode (Gaastra 1966*b*).

intensity as electrons are excited from the inactive, bound state to become free electrons, decreasing the resistance of the material.

(*c*) The *bimetallic* actinograph exploits the distortion of a blackened bimetallic strip exposed to radiation.

(*d*) In a Bellani pyranometer, radiant energy is measured by the amount of *liquid distilled* from a blackened copper sphere into the graduated glass tube attached to it.

(*e*) *Photochemical* methods were often recommended in the past, but it is very difficult to establish any consistent relation between these and other measurements. Even if their strong spectral bias were eliminated, it would not be easy to construct a container with suitable directional properties.

MEASUREMENT OF RADIATION ABOVE THE CANOPY

Solar radiation (0·3–3·0 μ)

Solarimeters

These are thermopile instruments whose blackened surfaces are uniformly sensitive to all wavelengths. The sensing surface is protected from wind and rain by glass domes which restrict the spectral sensitivity to 0·3–3·0 μ. When turned to face the ground these instruments measure the reflected short-wave radiation, but caution is necessary when solarimeters with large domes are inverted because the internal convective heat transfer is upset and sensitivity increases. Because their receiving surface is not shiny, they have a good directional response, provided the glass domes have a uniform thickness.

Two very good commercial instruments are: the Kipp solarimeter made in Holland, and the Eppley pyranometer made in the USA. Some cheaper instruments are on sale or can be made in the laboratory.

Kipp solarimeter. This instrument employs a Moll type of thermopile of 14 very thin constantan and manganin strips. Although it is a very stable and reliable instrument, it is not well suited for field work because of its weight. Kipp solarimeters are used for the continuous recording of solar radiation in Britain and elsewhere in Europe, and in Africa.

Eppley pyranometer. This instrument uses a gold-palladium and platinum-rhodium thermopile to measure the temperature difference between two concentric black and white rings. The sensing element is a disc, about 2·8 cm in diameter and 0·5 cm in height, enclosed in a 7·5 cm diameter glass bulb. It is lighter than the Kipp, more expensive, and less robust.

These solarimeters are too costly to use in large numbers, and also too heavy or fragile, but cheap units can be bought or can be made in the laboratory workshop.

The *Stern pyranometer*, made in Austria, uses a thermopile with the black and white segments alternating radially. (Dirmhirn 1954).

A solarimeter made by *Lintronic Ltd.*, London in co-operation with Rothamsted has an 80 junction copper-constantan thermopile made by the printed circuit technique, deposited on a thin fibre glass sheet in two concentric circles. The smaller one on the inside is covered by a central black spot (hot junction) and the larger, outer circle (cold junction) is placed under a white annulus. This instrument—like the Eppley—measures the temperature difference between the black and white areas.

The *Rothamsted type* of solarimeter has a circular, toroidally wound thermopile of copper-constantan junctions, made by electroplating copper onto bare constantan wire. The hot junctions are arranged on the central black spot and the opposing cold junctions on the white annulus (Monteith 1959).

Table 3 summarizes some of the salient features of these solarimeters.

TABLE 3

Solarimeter	Approx. output (mV/cal cm^{-2} min^{-1})	Approx. int. res. (ohms)	Dome	Approx. weight (kg)	Response-time (sec)*
Kipp	8	8	Optically ground	3	?
Eppley	7	100	Optically ground	1	4
Stern	2	5	Not ground	0·20	2
Lintronic	15	90	Not ground	0·10	10
Rothamsted	5	13	Not ground	0·03	20

* Time to reach $E/E_{max} = 1 - 1/e$ (63 per cent of full reading).

Photocells and solar cells

Selenium cells have been used extensively in ecology because they are cheap and easy to handle, but in experiments when the quality of radiation changes (e.g. inside the canopy) or when different radiation sources are compared (e.g. daylight and artificial light) these cells give misleading information (see comparison on p. 114 and Table 1.) However, to measure the incoming solar radiation in less demanding work, photoelectric cells are useful provided they are calibrated against a thermopile instrument (Trickett & Moulsley 1956). Calibration is possible because the spectral composition of total solar radiation does not change much with the sun's height or the cloudiness. Silicon solar cells are preferable to selenium because they are more stable. Selenium cells have inaccurate cosine response so they are usually exposed under cosine-corrected diffusers. Silicon cells are mounted without diffusers, sometimes under glass domes, but because of their shiny surface, they too are less accurate at low solar elevation. Photo-tubes and photo-resistors, with the notable exception of the 'Kubin' light integrator (see 'Visible radiation', p. 120), are very rarely used to measure radiation.

Bimetallic actinograph

The Casella bimetallic actinograph developed at Kew Observatory from the original Robitzsch instrument works mechanically by recording the difference in expansion in three bimetallic strips, two of which are shielded, and the third blackened and exposed to radiation. The two shielded strips, fixed to the frame, compensate for changes in air temperature. The bending of the third blackened strip is transmitted to the chart on a clockwork drum. Although it gives a chart record only daily totals of radiation are likely to be accurate because it responds very slowly to changing radiation intensities. Careful adjustment and maintenance are essential (Blackwell 1953).

Gunn Bellani radiation integrator

In the Gunn-Bellani instrument, radiation is measured by the amount of liquid (water or alcohol) distilled from a blackened copper sphere into a graduated burette. In one model, the space in the sphere is evacuated to about 10–15 mmHg to lower the boiling point of the enclosed liquid. The copper sphere is surrounded by a glass envelope and the space between is evacuated to minimize convective heat transfer. The glass burette joined to the sphere is about 60 cm long. When the instrument is exposed, sunk in the ground with the sphere just protruding, the burette is always cooler and functions as a condenser. This instrument can give only integrated totals of radiation. The sensitivity depends on the ambient temperature and hence changes with season (Monteith & Szeicz 1960). A careful comparison with a thermopile integrator where it is to be exposed is absolutely essential.

Both the bimetallic actinograph and the Gunn-Bellani integrator are attractive because they are simple, do not need a power supply and can be left unattended for days at remote sites. These advantages must be offset against the considerably greater accuracy that can be achieved with cheap solarimeters, or with silicon solar cells operating small electrolytic or battery-powered transistorized integrators.

The use of photochemical methods, which in the main respond only to the ultraviolet part of the solar spectrum, is very difficult to justify today when much more accurate instruments can be made or bought.

Visible (0·4–0·7 μ) and infra-red (0·7–3·0 μ) radiation

Because plants use energy for photosynthesis in the 0·4–0·7 μ band, it is of special interest to discover how much of the solar radiation is in this waveband. In measurements, the solar spectrum could be separated either by excluding the infra-red beyond 0·7 μ, or by transmitting only the infra-red and obtaining the visible component by difference. There is no filter with a sharp cut-off above 0·7 μ, so the second approach is used. The Schott RG8 filter glass is ideal because it cuts off radiation sharply below 0·7 μ. It is available as a disc 5·0 cm diameter and 0·3 cm thick.

This filter mounted over a small 4 cm diameter solarimeter can be used for spot measurements on the visible energy content of any radiation source. First, the total short-wave radiation intensity is measured with a Pyrex glass disc mounted over the solarimeter on a tubular spacer. This compensates for the cut-off of radiation by the sides of the tube holding the filter and also for the losses by reflection. Then the RG8 filter is substituted and the difference between the readings gives the energy content of the source in the 0·4–0·7 μ region.

This arrangement is adequate for occasional measurements with an overhead source, but outdoor measurements with the sun at different angles during the course of the day need a different instrument. In the Rothamsted type band-pass solarimeter (Szeicz 1966), the RG8 and Pyrex filters are mounted 0·3 cm above a redesigned thermopile and covered with glass domes to protect them from the weather. When glass filters are used, they absorb radiation and heat up, so that the zero changes. The redesigned thermopile has a radial symmetry that eliminates this fault. The two receivers, one with Pyrex glass measuring total radiation, the other with RG8 measuring the infra-red component, are mounted side by side. Exposed for a year at Cambridge, the ratio of visible to total radiation varied slightly between 44 and 47 per cent on days with average cloudiness, but on completely overcast occasions it was more, nearing 50 per cent. For analysis of photosynthetic efficiency of crops, existing records of solar radiation can safely be used by assuming that 45 per cent of the total energy is in the visible range.

A similar but very expensive instrument using precise wire-wound thermopiles and optically ground domes of different filter glasses (Schott GG14, OG1, RG2, and RG8) is made by the Eppley Laboratory in the USA.

The 'Kubin' radiation integrator, made in Czechoslovakia, uses a photo-emissive tube whose spectral sensitivity is corrected by selectively absorbing glass filters (Kubin & Hladek 1963). The resulting combined spectral sensitivity never varies more than ± 5 per cent in the 0·37–0·73 μ region in which the instrument is sensitive. An added advantage is that the output is recorded on an integrating counter.

Net radiation (0·3–100 μ)

Net radiation is much more difficult to measure than solar radiation, and it is only during the last 10 or 15 years that reliable and accurate instruments have been designed for field use. The difficulty arises because net radiation is often a rather delicate balance between two large opposing fluxes of radiation in all wavelengths. Generally, thermopiles have to be arranged to measure the temperature difference between the top and bottom surfaces of blackened discs or plates exposed to radiation. With an uncovered plate in the free air almost all the absorbed radiation will be dissipated by free or forced convection. Since the convective heat transfer is proportional to the product of the square root of the wind-speed and the temperature difference between the surface of the plate and the air, the temperature difference *across the plate*, and hence the output of the thermopile per unit net radiation, will also depend on wind-speed (Gier & Dunkle 1951). Protection is difficult, because except for a thin polythene film, and a rare and expensive salt—thallium iodide (KRS 5)—no materials that could be used in the field transmit satisfactorily at all wavelengths.

Two arrangements can be used to overcome this difficulty. One is to ventilate the flux-plate with a fan fast enough (10–15 m sec^{-1}) to swamp the effect of the wind. The second is to cover both top and bottom surfaces with thin polythene hemispheres and pass dry air or nitrogen through under slight pressure to prevent collapse and condensation.

In less ambitious laboratory-made instruments, flat polythene film can be used. Another way to decrease the wind-sensitivity is by using a material with high thermal conductivity (crown glass, or anodized aluminium). With this, the temperature difference *across* the plate will be rather small, resulting in much more equal convective heat loss from both top and bottom surfaces.

Ventilated instruments

All ventilated net radiometers originate from the Gier & Dunkle design of 1951. A blackened flux-plate of about 10 × 10 cm is mounted in a nozzle which in turn is fixed to a large and heavy blower housing. As made by the Beckman and Whitley Co. in the USA, the reading of the plate can be disturbed by the heat radiated by the electric motor operating the fan. Some workers have modified the design by using a smaller and faster fan and a much longer nozzle. An interesting feature of one of the modifications (Suomi, Fransilla & Islitzer 1954) is a small air-foil built into the nozzle. Varying the angle of the air-foil provides a sensitive control for balancing the ventilation over the plate (see p. 128).

It is easy to make this type of radiometer in the laboratory because fans usually employed in air conditioning are readily available. The nozzle can be made of Perspex or Bakelite, and the flux-plates by electroplating.

The disadvantages are that it is rather heavy and requires power supply to operate continuously. Because the flux-plate is not protected, its reading in rain is meaningless, but this is not a serious drawback because below overcast skies the net long-wave exchange is negligible and net radiation can be estimated from the short-wave component alone.

Polythene-shielded instruments

Schulze's (1953) net radiometer was the first to use a polythene wind-shield. In principle it is like two Kipp solarimeters joined together, one facing up, the other down. The glass domes are replaced by stiff polythene hemispheres, and dry air is circulated to avoid condensation. In conjunction with two short-wave instruments, it was used originally to measure the individual components of the radiation balance. Because the temperature of the two instrument casings must be measured separately, a recorder with four channels is required. Although excellent as a research tool in radiation meteorology, it is too cumbersome and complicated for general field use.

The Funk net radiometer uses a specially made 'ribbon' thermopile (Funk 1959). A narrow strip of thin plastic is wound with constantan wire and copper-plated so that the thermo-junctions are situated on the edge of the strip. When this is rolled into a loose spiral, the junctions are closely packed. It is mounted in a metal heat sink and thin aluminium foils are cemented in good thermal contact onto the junctions. Very thin (0·01 mm) moulded polythene hemispheres shield the blackened faces from wind and rain, and are kept inflated and dry by slowly passing nitrogen through the instrument. The output of the instrument is around 24 mV/cal cm^{-2} min^{-1}, with an internal resistance of about 100 ohms. It is a high grade instrument manufactured by Middleton Pty in Australia. Accessories will convert the instrument to measure the short-wave balance or the incoming short and long-wave radiation. Several variants of the polythene shielded instrument can be made in the laboratory (Fritschen 1965).

An unshielded and unventilated instrument described by Monteith & Szeicz (1962) consists of a glass ring 10·2 cm outer and 7·6 cm inner diameter, 0.15 cm thick, wound tightly with constantan wire (a 32 s.w.g. wire was used first but 42 s.w.g. is better.) The wire is electroplated to form copper–constantan junctions on the top and bottom surfaces. Use of a crown glass ring and thin wire wound in close contact with it ensures that the output is almost independent of windspeed. Winding on 280 turns, the instrument gave 9·2 mV/cal cm^{-2} min^{-1}, with an internal resistance of 200 ohms. Wind sensitivity in the range of 0·5–3·5 m sec^{-1}, most frequently encountered on the field, was $\pm$2·5 per cent. Requiring no power or nitrogen supply, the instrument gives an output accurate enough to provide daily or weekly totals of net radiation when integrated.

MEASUREMENTS OF RADIATION WITHIN THE CANOPY

Solar radiation (0·3–3·0 μ)

A question often asked in crop ecology is how much solar radiation is transmitted to a specified level in a crop. Conventional solarimeters are not suitable for measurement below the canopy because the radiation is rather uneven. To overcome this, a tube solarimeter was designed at Rothamsted with a thermopile 2·5 cm wide, 90 cm long, enclosed in a glass tube (Szeicz, Monteith & Dos Santos 1964). When mounted among the plants, it averages radiation over an area of 225 cm^2. The black and white areas are arranged alternately to minimize the effect of heating on one side of the glass. Output is about 25 mV/cal cm^{-2} min^{-1}, with an internal resistance of 40–50 ohms. Several of these units are exposed on a sloping stand, and much of the undesirable directional response of the glass tube is overcome by using an identical unit above the stand with which all the others in the crop are compared. A miniature version, 0·8 cm in diameter, 35 cm long, was built to use in grass and herbage (Szeicz 1965). Its output is about 12–14 mV/cal cm^{-2} min^{-1}, with an internal resistance of 50–60 ohms.

Visible (0·40–0·75 μ) *and infra-red* (0·75–3·0 μ) *radiation*

A similar large unit fitted with a Wratten 88A gelatine filter, transmitting between 0·75 and 3·0 μ, measures the transmission of infra-red radiation. By difference between the clear and filtered units, the intensity of visible radiation between 0·4 and 0·75 μ can be found. Figure 3 shows clearly how visible radiation is selectively absorbed in a crop of barley and kale.

Net radiation (0·3–100 μ)

To study the exchange processes below the canopy, profiles of net radiation are needed. Ventilated instruments cannot be used because they would disturb the environment. Recently, standard polythene shielded instruments have been used successfully (Brown & Covey 1966), but longer units are needed for spatial averaging.

In this kind of instrument two methods will eliminate the need for ventilation: using a long strip-like flux-plate made from a material of high thermal conductivity (e.g. aluminium strip anodized or insulated with thin tape); or enclosing a long flux-plate in thin-walled polythene tubing (Burrage 1966).

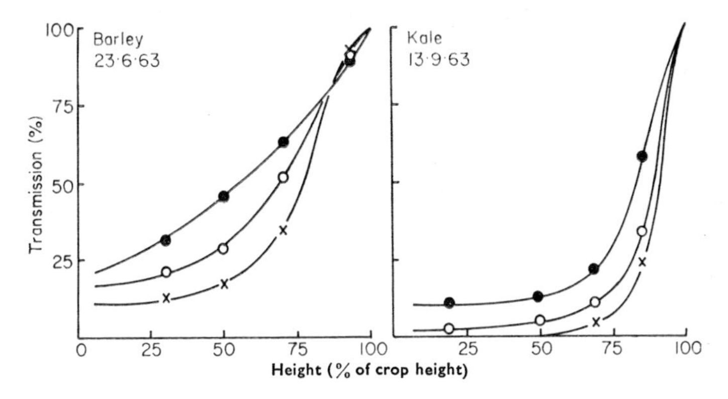

FIG. 3. Transmission of radiation in 2 crops; *a*, barley; *b*, kale. ●, Total radiation (0·4–2 μ); ○, infra-red (0·75–2 μ); ×, 'visible' (0·4–0·75 μ) (Szeicz, Monteith & Dos Santos 1964).

A model of the first type was made from an aluminium strip 12·5 cm long, 2·5 cm wide and 0·3 cm thick, covered with Sellotape and wound with 42 s.w.g. (0·01 cm diameter) constantan wire at 8 turns per cm. The wire was copper-plated, and gave a sensitivity of 1·7 mV/cal cm^{-2} min^{-1} per 10 cm length, with 45 ohms internal resistance per 10 cm. A 100 cm long unit, similarly wound, would give an output of 17 mV with an internal resistance of 450 ohms, but if needed, a unit with a smaller internal resistance could be made by sub-dividing the strip into smaller sections in parallel as in the tube solarimeter. The sensitivity of the prototype to wind is only ±2 per cent in the critical range below 1 m sec^{-1}. With this type of sensor continually exposed in the canopy, one foreseeable difficulty would be the dust and dew depositing on the surface.

Alternatively, a long net-radiometer shielded by polythene may be better but dust and dew may still influence the reading. A net radiometer of this kind is now made in Australia by Swissteco Pty of Melbourne.

CALIBRATION AND USE OF INSTRUMENTS

Precise instruments sold by reputable manufacturers carry a calibration certificate, but because the sensitivity can change suddenly, as a result of shock in transit, or can drift slowly while the instrument is in use, it is always advisable to check this at one of the recognized radiation stations.

Solarimeters. The high grade instruments continuously recording solar radiation should be checked, preferably every year, against substandards.

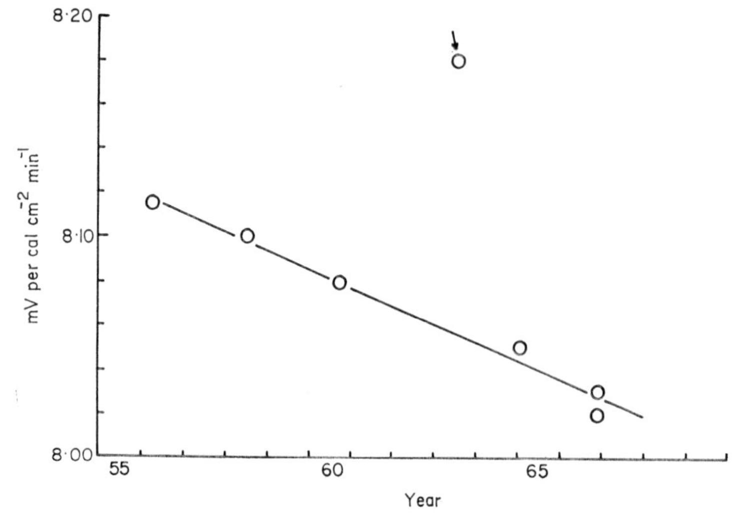

FIG. 4. Change in sensitivity of the Kipp solarimeter, exposed at Rothamsted since 1956.

The sensitivity of these instruments is usually very stable, but thermopile efficiency or blackening may deteriorate gradually. Figure 4 shows how the sensitivity of the Kipp solarimeter exposed at Rothamsted changed since 1956. These calibrations were done either at NIAE Silsoe or at Kew Observatory. The very slight decrease in sensitivity is consistent and amounts to only 0·1 per cent per year. The 1963 point, obviously not correct, is included to show how readily a wrong calibration can be detected retrospectively.

With continuously recording instruments it is very convenient to have the scale of the recorder graduated in radiation units. Even better, radiation can be integrated by the recorder and totals shown on a counter, or printed at intervals.

When calibrating a secondary instrument, the best method is to use a diffusing sphere (Fig. 5), or compare the instruments in the open, because sharply focused light gives false readings with optically imperfect domes. When there is a spare recorder or recorder channel, the two instruments can be exposed side by side for a few days, half-hourly or hourly means

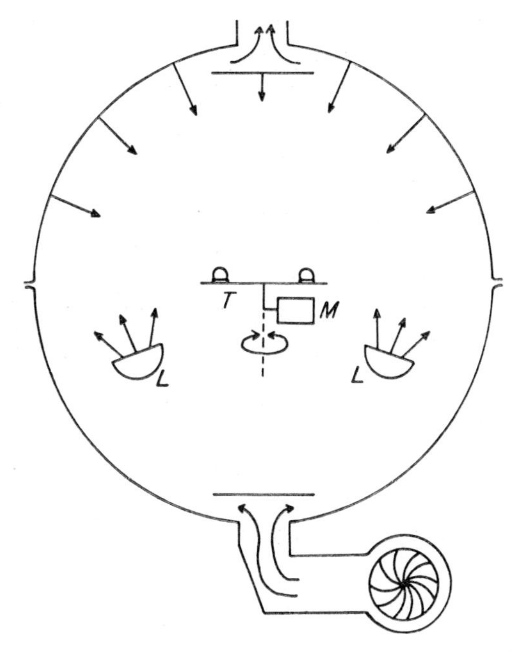

FIG. 5. An integrating sphere, used to calibrate solarimeters. The sphere is about 2 m in diameter, painted matt white inside. Air is circulated to prevent the temperature rising. The lights *L* are evenly spaced in respect to the turntable *T* so that illumination is uniform. The turntable with the solarimeters is slowly oscillated through about 360° by the motor *M* once approximately ever 5–6 min.

plotted against each other, and the line of best fit determined to give the calibration factor. Alternatively, with no recording facilities available, in steady conditions the output of the secondary instrument can be com-compared with the standard at several different radiation intensities, and the mean calibration factor calculated. Solar cells, whose cosine response is bad at low elevations, should be calibrated only when the sun's elevation is not less than about 40°. Readings will be suspect when using these cells with low sun.

To calibrate solarimeters and solar cells with integrators, they should be run for at least a week, and half-daily or daily totals plotted against the readings from the standard solarimeter to determine the calibration factor.

Some secondary instruments are not very stable, and it is important to check their calibration frequently, for example, before and after exposure for a few months in the field, and more often when the instruments are subjected to rough handling.

If many instruments are used, it is sound policy to keep one as a standard to check others. Too often a long run of measurements, collected at great effort, has to be discarded because of uncertain calibration.

When using several instruments at one site to compare radiation intensities (e.g. when measuring transmission of radiation in crops, in water, in glasshouses, or measuring the reflection coefficient), it is often unnecessary to calibrate the instruments in absolute units. Comparison is facilitated by balancing the outputs of receivers with different sensitivities, by exposing them side by side, carefully levelled on a day with steady radiation.

With two instruments only, a 'ratio bridge' arrangement (Monteith & Szeicz 1962) accelerates measurements in quickly changing radiation.

Band-pass solarimeters, using cut-off filters below $0 \cdot 75\,\mu$, can be calibrated under a curved sheet of Wratten 88A gelatine filter, which transmits only radiation to which those instruments fitted with the filters are equally sensitive.

Net radiometers. Net radiometers are less accurate than solarimeters because they must measure a small difference between two large fluxes, and at the same time have to respond to short and long-wave radiation equally. In ventilated instruments, equal response is ensured by the uniform black sensor and by the absence of a selectively absorbing cover, but balanced ventilation is more difficult to achieve. When the ventilation is not balanced the two sides have different sensitivites, and the output is not proportional to net radiation. To test it for balance the instrument is exposed to a constant radiation intensity and then inverted: the two outputs should not differ by more than 2 or 3 per cent.

When assembling polythene shielded instruments, care must be taken that the thermopile covers for both the top and bottom surfaces are in good contact ensuring thermal symmetry. To balance short and long-wave sensitivities after transmission through polythene—which absorbs some of the long wave radiation—about 5 per cent of the sensing surface is painted white as on the Funk instrument.

After balancing for symmetry and wavelength, a net radiometer can be calibrated in two ways. In the laboratory under a lamp, two separate sheets of glass must be interposed between the lamp and the instrument, the first sheet absorbing all the long-wave radiation emitted by the lamp, and the second absorbing the long-wave radiation emitted by the first sheet of glass. Thus the radiation transmitted by the two sheets will be the short-wave component only. Two more sheets below the instrument serve the same purpose for radiation reflected and emitted from below. If the underlying surface is black, then the instrument will 'see' only the incoming short-wave component, whose intensity can be measured with a solarimeter, and the calibration factor calculated.

Outdoors, calibration under cloudless or nearly cloudless skies is simpler. A solarimeter and the net radiometer are exposed so that the sensing surfaces are shaded from direct sunshine. When steady readings are reached, the shades are removed. The differences between the shaded and unshaded readings of the solarimeter and the net radiometer are compared, and the mean calibration factor calculated. This is relatively easy if the output of both instruments is recorded on a pen recorder, but, if not three pairs of hands are needed. Once a first class instrument of either a Gier & Dunkle type or a Funk type is calibrated other net radiometers can be calibrated directly against it.

Net radiation is best recorded continuously and, provided the cost is justified, can be integrated by the recorder. It is also possible to integrate net radiation electrically, but because the signal is negative overnight, a biasing voltage must be provided, larger than the maximum negative signal that can occur on a clear night.

REFERENCES

BLACKWELL M.J. (1953) On the development of an improved Robitzsch-type actinometer. *Met. Res. MRP.* No 791, 10 pp.

BLACKWELL M.J. (1966) Radiation meteorology in relation to field work. In: *Light as an Ecological Factor.* (Ed. by R.Bainbridge, G.C.Evans & O.Rackham), p. 17–41. Blackwell Sci. Publ., Oxford.

BROWN K.W. & COVEY W. (1966) The energy budget evaluation of the micrometeorological transfer processes within a cornfield. *Agric. Met.* **3**, 72–96.

BURRAGE S.W. (1966) In: *Investigation of Energy, Momentum and Mass Transfer near the Ground.* University of California, Davis. Final report 1965. Task number: DA Task 1 VO-14501-B53A-08 p. 65–7.

CALLAHAN P.S. (1965) Intermediate and far infra-red sensing of nocturnal insects. *Ann. ent. Soc. Am.* **58**, 727–56.

DIRMHIRN I. (1954) Einfache Sternpyranometer. *Wett. Leben*, **6**, 41–6.

FEDERER C.A. & TANNER C.B. (1965) A simple integrating pyranometer for measuring daily solar radiation. J. *geophys. Res.* **70**, 2301–6.

FRITSCHEN L.J. (1965) Miniature net radiometer improvements. *J. appl. Met.* **4**, 528–32.

FUNK J.R. (1959) Improved polythene shielded net radiometer. *J. scient. Instrum.* **36**, 267–70.

GAASTRA P. (1966*a*) Some comparisons between radiation in growth-rooms and radiation under natural conditions. *Proc. 1st UNESCO Symp. on Ecosystems, Copenhagen.* In press.

GAASTRA P. (1966*b*) Radiation measurements for investigation of photosynthesis under natural conditions. *Proc. 1st UNESCO Symp. on Ecosystems, Copenhagen.* In press.

GIER J.T. & DUNKLE R.V. (1951) Total hemispherical radiometers. *Trans. Am. Inst. elect. Engrs*, **70**, 339.

KUBIN S. & HLADEK L. (1963) An integrating recorder for photosynthetically active radiant energy with improved resolution. *Pl. Cell Physiol., Tokyo*, **4**, 153–68.

MONTEITH J.L. (1959) Solarimeter for field use. *J. scient. Instrum.* **32**, 341–6.

MONTEITH J.L. & SZEICZ G. (1960) The performance of a Gunn-Bellani radiation integrator. *Q. Jl R. met. Soc.* **86**, 91.

MONTEITH J.L. & SZEICZ G. (1962) Simple devices for radiation measurements and integration. *Arch. Met. Geophys. Bioklim.* B **11**, 491–500.

SCHULZE R. (1953) Uber ein Strahlungsmessgeret mit ultrarotdurchlassiger Windschutzhaube am Meteorologischen Observatorium Hamburg. *Geofis. pura appl.* **24**, 167.

Smithsonian Meteorological Tables (1951) Sixth Revised Edition. *Smithsonian Miscellaneous Collections.* Vol. 114. Washington, D.C. pp. 413 and 415.

SUOMI V.E., FRANSILLA M. & ISLITZER N.F. (1954) An improved net-radiation instrument. *J. Met.* **11**, 276–82.

SZEICZ G. (1965) A miniature tube solarimeter. *J. appl. Ecol.* **2**, 145–7.

SZEICZ G. (1966) Field measurements of energy in the 0·4–0·7 micron range. In: *Light as an Ecological Factor* (Ed. by R.Bainbridge, G.C.Evans, & O.Rackam), p. 41–51. Blackwell Sci. Publ., Oxford.

SZEICZ G., MONTEITH J.L. & DOS SANTOS J.M. (1964) Tube solarimeter to measure radiation among plants. *J. appl. Ecol.* **1**, 169–74.

TRICKETT E.S. & MOULSLEY L.J. (1956) An integrating photometer. *J. Agric. Engng Res.* **1**, 1–11.

THE MEASUREMENT OF CARBON DIOXIDE CONCENTRATION IN THE ATMOSPHERE

G.E.Bowman

National Institute of Agricultural Engineering, Silsoe

SUMMARY Carbon dioxide is now cheap enough to use as an aerial fertilizer under glass and so the grower needs simple equipment for gas analysis. The research worker often requires gas analysis equipment of high sensitivity and good stability. The salient features, including approximate cost, of the following methods of gas analysis are discussed: infra-red absorption, electrical conductivity of alkali solution, electrical conductivity of deionized water, equilibrium pH of alkali solution, titration of alkali solution, chemical absorption (volumetric), chemical absorption (indicator tube), gas-liquid chromatography, optical interferometry and thermal conductivity.

INTRODUCTION

Methods of measuring CO_2 concentration for ecological purposes can conveniently be divided into two groups: methods suitable for use in the field and laboratory methods. For field use, an instrument must be portable and easy to use, cheap to buy and cheap to operate. Since field measurements are at or near the normal atmospheric CO_2 concentration, very high sensitivity is unnecessary. For laboratory use, higher standards of sensitivity and stability of calibration are required. Further important considerations are air sample size, the response time of the apparatus and whether or not the apparatus can be readily adapted for the control of CO_2 concentration,

Since CO_2 is only a minor constituent of the atmosphere, concentrations of the gas are more conveniently expressed in terms of volumes of CO_2 per million volumes of air (abbreviated to vpm), rather than as a percentage.

FIELD METHODS

1 *Optical interferometer*

This instrument detects changes in the refractive index of a column of air. The refractive index of air at 0°C is 1·00029, whereas that of CO_2 is 1·00045 at the same temperature. Changes in refractive index are indicated by the movement of a system of interference fringes which are observed through an eyepiece. The distance a particular fringe moves is measured against a scale which is connected to a vernier adjustment. True zero setting can be achieved by filling the instrument with CO_2 free air. An accuracy of 60 vpm is claimed, provided that temperature and barometric pressure corrections are applied to the results. The instrument is non-selective and moisture must be removed from the air sample by means of a drying tube. Although widely used in the Netherlands by horticulturalists it is not considered at the N.I.A.E. to be an entirely satisfactory instrument for use in glasshouses. Considerable practice in operation is needed before repeatable results can be obtained.

2 *Chemical absorption (indicator tube)*

Three instruments of this type are available, all of which use a pump to draw an air sample through a detector tube containing a CO_2 absorbent chemical and an indicator. When a fixed volume of air containing CO_2 is passed through the detector tube a colour change is induced in the indicator; the length of the stain so produced is a measure of the CO_2 concentration in the air sample. In one example of an instrument of this type (Plate 1a), a finely machined hand pump is used to draw 100 ml of air through the detector tube which is 10 cm long and contains an indicator absorbed on fine-grain silica gel. The colour change in the presence of CO_2 is from blue/purple to pale pink. The length of the colour stain is related to the CO_2 concentration by means of a calibration chart which is provided with each carton of 20 tubes. Temperature corrections are given on the calibration chart and cover the range 0–40°C. For one stroke of the pump, the range is 0–7000 vpm CO_2 and for a sample of 300 ml, i.e. three strokes of the pump, the range is 0–1500 vpm CO_2. This type of instrument is selective and detector tubes are available for measuring a wide range of both organic and inorganic gases. Tubes for measuring CO_2 cost about five shillings each. Use may be made of both ends of a tube if (a) the first stain does not extend beyond a third

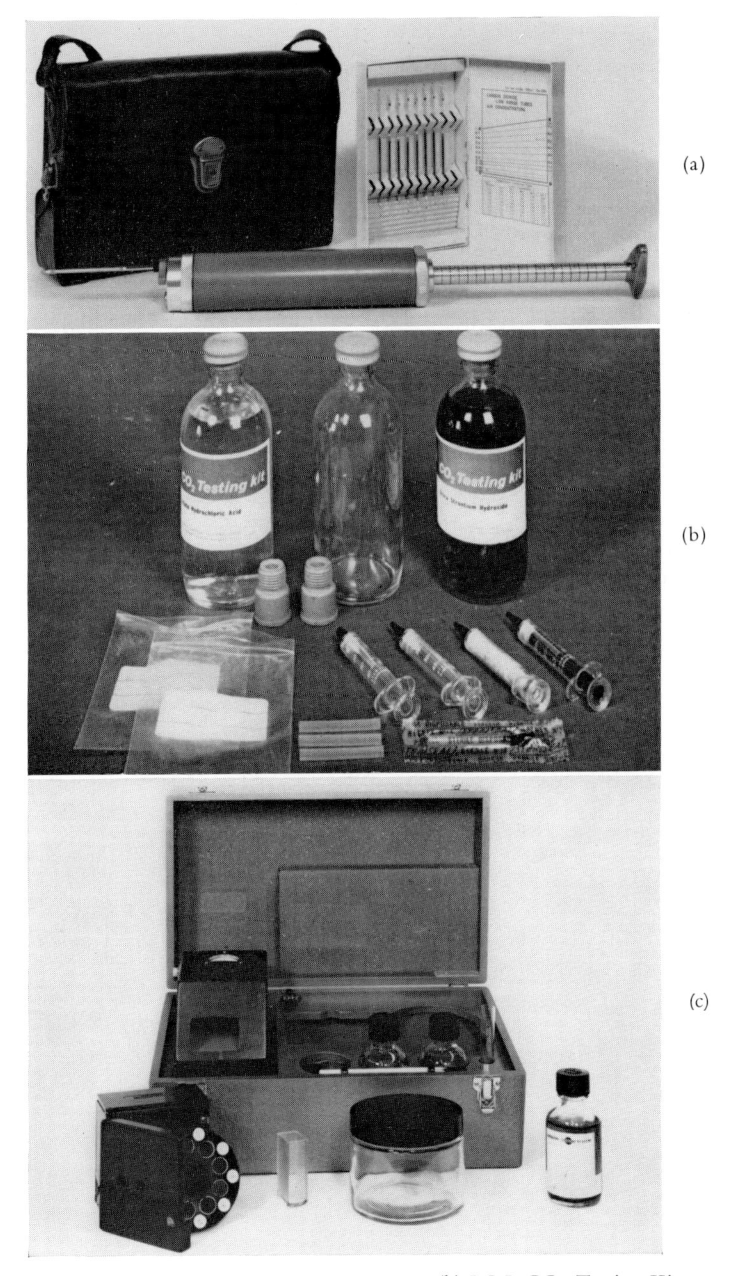

PLATE 1. (a) Kitagawa Precision Gas Detector; (b) I.C.I. CO_2 Testing Kit; (c) Tintometer CO_2 Kit.

[*facing p.* 132

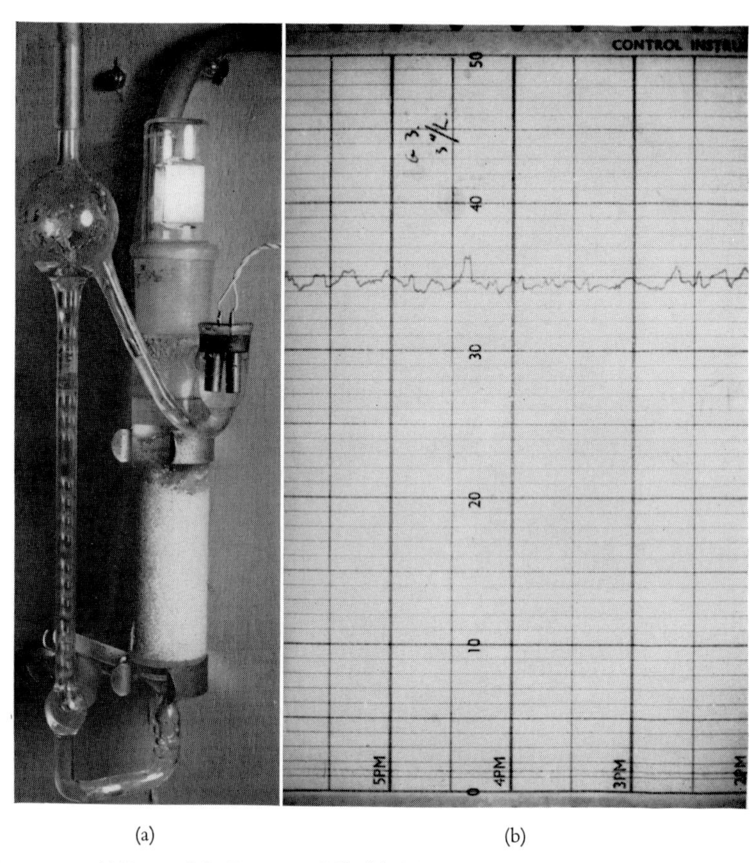

(a) (b)

PLATE 2. (a) James CO_2 Detector Cell; (b) CO_2 concentration control in a glasshouse using the James cell and an N.I.A.E. control circuit (1 division = 23 vpm CO_2).

of the filled portion (b) the second measurement is completed within a few minutes of the first.

3 *Titration of alkali solution*

In this method the CO_2 from a sample of air is absorbed in a dilute solution of strontium hydroxide containing a blue indicator. Aliquot amounts of dilute hydrochloric acid are then added until the indicator becomes colourless, the quantity of acid added being a measure of the CO_2 concentration in the sample. An air sample is taken by emptying a 250 ml bottle, previously filled with distilled water, at the sampling point. The reagent and sampling bottles are stoppered with serum caps through which the reagents may be extracted or injected with the aid of simple 2 ml syringes. Separate syringes must be used for the acid and for the alkali. If the CO_2 concentration is so high that the indicator becomes colourless before the addition of any acid, a further 2 ml of alkali is added; this is equivalent to 900 vpm CO_2. Each 0·2 ml of acid is equivalent to 90 vpm CO_2. The method tends to underestimate the CO_2 concentration and it is sometimes difficult to avoid the formation of bubbles in the syringes. The apparatus is shown in Plate 1b.

The titration of potassium hydroxide was used by Heinicke and Hoffman (1933) in measurements of the photosynthesis of apple leaves.

4 *Visual comparative colorimetry*

This technique has been developed at the N.I.A.E. by Sharp (1964), based on the methods of Claypool & Keefer (1942) and of Čatský & Slavíck (1960). The pH of a solution of 0·001 N sodium bicarbonate and 0·1 N potassium chloride is determined colorimetrically. The pH, indicated by a mixture of cresol red and thymol blue, is directly related to the equilibrium partial pressure of the CO_2 in the air in contact with the solution. A wide-mouthed screw-topped jar containing 10 ml of solution is left at the sampling point, with the top removed, for a period of 30 minutes. The top is replaced whilst the jar is taken back to the laboratory where the solution is poured into a rectangular glass viewing cell. The colour of the contents of the cell is then compared against a comparator disc. The disc has nine different glass colour standards selected to represent the following equilibrium CO_2 concentrations: 200, 300, 400, 500, 750, 1000, 1500, 2000 and 4000 vpm. The corresponding colour changes are from deep red, through orange, to bright yellow. For viewing the colours, a battery-operated white light source forms part of the apparatus and it is essential to use this, since the spectral transmissions of the glass colour standards and of the indicator solution are not identical, but were matched using a standard white light source. Some men have defects in colour vision and may therefore have difficulty in matching the colours. A thermometer and temperature correction chart, covering the range 0–30°C, are included in the equipment. Problems have arisen with the storage of the solution which so far has been supplied in 100 ml bottles. In future it is proposed to supply concentrated reagent in small sealed glass ampoules, deionized water being added just before use. Attention to cleanliness is important; it is recommended that the sampling jar and viewing cell are rinsed with a little fresh indicator solution, or solution remaining from a previous measurement, before use. The apparatus is shown in Plate 1c.

LABORATORY METHODS

5 *Katharometer*

Changes in CO_2 concentration are detected by measuring the thermal conductivity of the gas mixture. At 20°C, the thermal conductivity of air is $6{\cdot}7 \times 10^{-5}$ cal cm/cm^2 sec °C, whereas that of pure CO_2 at the same temperature is $4{\cdot}4 \times 10^{-5}$ cal cm/cm^2 sec °C. Thermal conductivity is deduced from the temperature of a heated wire over which the air sample is pumped. Robust instruments of this type are commonly used for boiler flue gas analysis where the working range is 0–20 per cent CO_2 and great sensitivity is not required. The smallest change in CO_2 concentration which can be detected by a hot-wire katharometer is about 200 vpm. The katharometer is easily adapted for control purposes.

6 *Chemical absorption (volumetric)*

The components of a gas mixture are absorbed in sequence by specific chemical reagents and the volume of gas remaining at each stage in the process is measured. A portable apparatus of this type for the measurement of CO_2 and other gases in mine air has been described by Haldane (1920). The method is slow and requires careful manipulation. A more sensitive form of this apparatus has been developed by Carpenter, Fox and Sereque (1929) in which changes in CO_2 concentration of 10 vpm can be detected.

7 *Electrical conductivity of alkali solution*

The electrical conductivity of a suitable alkaline absorption solution, e.g. sodium hydroxide for the detection of CO_2, is measured before and after the gas component has been absorbed. This method is much more sensitive than any of those considered so far and about 2 vpm CO_2 can be detected. Electrolytes have a large negative temperature coefficient of resistance, of the order of 2 per cent per °C. The principal object in measuring electrolyte resistance before and after the gas has been absorbed is to provide temperature compensation. The electrical parts of an instrument of this type are simple and automatic control is easily arranged. Practical disadvantages are that the flow rates of both gas and liquid through the apparatus must be maintained constant and frequent replenishment of the sodium hydroxide is necessary. The sodium hydroxide must also be made up accurately to a specified concentration. This method was used by Clark, Shafer & Curtis (1941) and also by Leach, Moir & Batho (1944). Steiner (1961) describes a commercially made instrument in which temperature correction is by means of thermistors.

8 *Electrical conductivity of deionized water*

Some three years ago a novel method of locating earthquake and avalanche survivors was proposed: the method was based on the use of deionized water through which an air sample was bubbled (James 1964). The CO_2 concentration in exhaled air is about six times higher than that of the atmosphere, sufficient to cause a large increase in the electrical conductivity of deionized water. The patentee of a CO_2 detector of this type (James 1966) made available to N.I.A.E. a prototype unit around which a controller suitable for use in commercial glasshouses has been built. The CO_2 detector consists of an air-lift pump connected to a mixed-bed column of ion exchange resins (Plate 2a, facing p. 133). Air is pumped into deionized water at the base of a vertical tube; as air bubbles ascend the tube some CO_2 is dissolved in the entrained water. At the top of the tube a separator allows the air to escape to waste and the water to flow through a conductivity cell. Having passed through the conductivity cell, the water then runs down the resin column to complete the circulation. A constant head device containing thin, non-volatile oil ensures the production of a regular train of bubbles.

The stainless steel electrodes of the conductivity cell are supplied with alternating current at 50 Hz and the voltage developed across a resistor in series with the cell provides a signal which is amplified, rectified and used to operate a trigger circuit. The trigger circuit operates a relay which energises a solenoid valve through which CO_2 is admitted to the glasshouse. Means are provided for adjusting the voltage at which the trigger operates to enable the controlled CO_2 concentration to be set to a chosen value. Because the correct operation of the controller depends upon the maintenance of a stream of bubbles in the detector, monitoring of the flow of bubbles is necessary, particularly so should the flow cease whilst the solenoid valve is energized and gas is being injected. A small lamp and photo-transistor are therefore placed on opposite sides of the bubble tube so that they occupy conjugate foci of the cylindrical lens formed by water in the tube. The passage of a bubble de-focuses the light and causes a reduction in photo-transistor current. The photo-transistor is connected to a frequency-sensitive circuit which energizes a relay only when bubbles pass at a suitable rate. The contacts on this relay are in series with those on the relay operated by the trigger circuit. If the lamp fails, the solenoid valve cannot be opened.

Plate 2b (facing p. 133) shows a chart record obtained with an infra-red absorption gas analyser of the performance of the controller set to 800 vpm CO_2 in a glasshouse of 70 m^3. For an air sample of constant CO_2 concentration, the cell conductivity depends on flow rate since the ratio of air to water in the ascending mixture and the transit times of individual bubbles depend upon flow rate. Experimentally, these effects are small; at a sample flow rate of 0·5 ml/sec a 10 per cent change in flow rate causes only a 1 per cent change in cell resistance. Compensation for changes in temperature by a method similar to that used for the electrical conductivity of an alkali solution cannot be used in this apparatus because the temperature coefficient of resistance of deionized water is not sufficiently well-defined. It would, however, be possible to provide temperature compensation by means of a thermistor immersed in the water in the conductivity cell.

The apparatus described is an experimental prototype—it is anticipated that commercially made equipment, suitable for use in growth chambers and glasshouses, will soon be available. A practical advantage of this method over the previous one is that the resin column requires replacement at infrequent intervals of the order of one month or more, depending upon the CO_2 concentration control level.

9 *Gas chromatography*

For applications where the minimum possible disturbance to the environment is required, gas chromatography is an attractive method of measuring CO_2 concentration. During some preliminary work on methods of CO_2 analysis at N.I.A.E., using components designed by Hill & Hook (1960) for a miniature portable chromatograph developed for the analysis of anaesthetic gas mixtures, it was found possible to measure changes in CO_2 concentration of 50 vpm with an air sample of only 1 ml. The chromatographic column consisted of triacitine absorbed on crushed firebrick (30–52 mesh) contained in a nylon tube of 2 mm bore and 3 m length. Hydrogen was used as the carrier gas and the elution time was 2 minutes. The detector was a katharometer incorporating thermistors.

The electrical output signal from a chromatograph is normally recorded on a strip chart recorder and the gas concentration measured is proportional to the area beneath the trace. Despite the fact that electronic integrators for this purpose are now available, it is still not an easy matter to adapt a gas chromatograph for gas concentration control.

10 *Infra-red absorption*

Infra-red absorption gas analysis offers very high sensitivity; if the air sample is compressed to ten atmospheres, then changes of 0·05 vpm CO_2 can be detected (Bartley 1964). Radiation from a dull red filament is reflected by mirrors into two tubes, one of which is a reference tube containing air from which the CO_2 has been removed. The air sample is pumped through the other tube. Radiation leaving the sample and reference tubes enters a detector consisting of two chambers filled with CO_2 and separated by a thin flexible metal diaphragm. Radiant energy within the CO_2 absorption spectrum not absorbed during its passage through the sample tube is absorbed by the CO_2 in the detector; differential thermal expansion of the gas causes movement of the diaphragm. Diaphragm movement is detected electrically and to facilitate electrical amplification, the radiation beams are chopped at about 7 Hz by means of a rotating shutter.

Water vapour absorbs strongly at many wavelengths in the infra-red waveband and interferes with the accurate measurement of CO_2. The use of drying tubes can be avoided by fitting lead telluride optical 'edge' filters to the detector cell windows, thus preventing the detection of wavelengths common both to water vapour and to CO_2.

Morris & Neale (1954) have described the use of an infra-red absorption instrument in a glasshouse. Infra-red analysis has been used by Gaastra (1959) for studies on the photosynthesis of single leaves, also by Monteith & Szeicz (1960) for the measurement of CO_2 flux over a field crop.

For the measurement of the net CO_2 assimilation rate of tomatoes in a glasshouse, infra-red analysers have been fitted with electronic controllers in which the rate of gas is proportional to the deviation of the measured gas concentration from the desired value (Bowman 1965).

II *Comparison of different methods of CO_2 analysis*

Table 1 summarizes the principal features of the methods described.

TABLE 1. A comparison of the methods described for measuring the concentration of CO_2 in air.

Method	Maximum sensitivity vpm	Sample size ml	Time for one determination min	Cost £	Control of CO_2 concentration
Infra-red absorption	0·05	1000	1	700	Yes
Electrical conductivity of alkali solution	2	100	0·5	500	Yes
Electrical conductivity of deionized water	10	50	1	200	Yes
Equilibrium pH of alkali solution	20	1000	30	20	No
Titration of alkali solution	50	300	5	2	No
Chemical absorption (indicator tube)	50	100	3	25	No
Chemical absorption (volumetric)	50	20	5	100	No
Gas chromatography	50	1	2	400	Difficult
Optical interferometer	60	100	0·5	170	No
Katharometer	200	200	2	150	Yes

REFERENCES

BARTLEY W.B. (1964) Non-dispersive infra-red instruments for process control. *Trans. Soc. Instrum. Technol.* **16**, 80–90.

BOWMAN G.E. (1965). In Press. The control of carbon dioxide in plant enclosures. *In:* F.E.Eckardt (Editor). Eco-Systems. *Proceedings of Copenhagen Symposium, UNESCO, Paris.*

CARPENTER T.M., FOX E.L. & SEREQUE A.F. (1929) The Carpenter form of the Haldane gas analysis apparatus. *J. biol. Chem.* **83**, 211–30.

ČATSKÝ J. & SLAVÍK B. (1960) A field apparatus for the determination of intensity of photosynthesis. *Biol. Plant.* **2**, 107–12. Russian text, Engl. summary.

CLARK D.G., SHAFER J. & CURTIS O.F. (1941). Automatic conductivity measuremenst of CO_2. *Plant Phys.* **16**, 643–6.

CLAYPOOL L.L. & KEEFER R.M. (1942) A colorimetric method for CO_2 determination in respiration studies. *Proc. Am. Soc. hort. Sci.* **40**, 177–86.

GAASTRA P. (1959) Photosynthesis of crop plants as influenced by light, carbon dioxide, temperature and stomatal diffusion resistance. *Meded. LandbHogesch. Wageningen* **59** (13) 1–68.

HALDANE J.S. (1920) *Methods of Air Analysis.* Griffin, London.

HEINICKE A.J. & HOFFMAN M.B. (1933). The rate of photosynthesis of apple leaves under natural conditions. Part 1. *Cornell Univ. Agric. Expt. Sta. Bull.* **577**.

HILL D.W. & HOOK J.R. (1960). Automatic gas-sampling device for gas chromatography. *J. Sci. Instrum.* **37**, 253–5.

JAMES D.B. (1964) Locating earthquake survivors. *Ion Exchange Progress* **3** (1), 1–4.

JAMES D.B. (1966) British Patent No. 1,018,658.

LEACH W., MOIR D.R. & BATHO H.F. (1944) An improved arrangement for the measurement of CO_2 output of respiring plant material by the electrical conductivity method. *Can. J. Res.* **22** Sec C, 133–42, Bibl. 11.

MONTEITH J.L. & SZEICZ G. (1960) The carbon-dioxide flux over a field of sugar-beet. *Quart. J. R. Met. Soc.* **86**, 205–14.

MORRIS L.G. & NEALE F.E. (1954) The infra-red carbon dioxide gas analyser and its use in glasshouse research. *N.I.A.E. Tech. Mem.* 99. Bibl. 7.

SHARP R.B. (1964) A simple colorimetric method for the in-situ measurement of carbon dioxide. *J. agric. Engng Res.* **9** (1), 87–94.

STEINER A.A. (1961) Het probleem koolzuurgas in de tuinbouw. *Jversl. Proefst. Groent. Fruitt.*, Naaldwijk, 120–5.

THE MEASUREMENT OF LIGHT AND TEMPERATURE AS FACTORS CONTROLLING THE SURFACE ACTIVITY OF THE SLUG *AGRIOLIMAX RETICULATUS* (MÜLLER)

P. F. Newell

Zoology Department, Westfield College, London N.W.3

SUMMARY Surface crawling normally occurs at night. Measurement of temperature changes in the soil using an 8-channel recorder and thermistor probes showed the gradual lowering of soil temperatures throughout the night, the lowest being recorded at the surface after daybreak. Nightfall and daybreak were recorded by a photoelectric cell connected to the recorder. Surface crawling, recorded photographically, appears to have been regulated by diurnal light changes.

INTRODUCTION

Slugs are normally found in the top 10 cm of the soil. Most of them feed on plant material and some are important agricultural pests (Howitt 1961).

Slugs can be separated from the soil by a flotation technique, which can be used for population estimation (South 1964), but this method is too laborious for normal field use. Other, less time-consuming, methods, that rely on trapping animals crawling on the soil surface, are inaccurate, mainly because they sample an unknown proportion of the population (Barnes & Weil 1942; Bett 1960; Webley 1964).

On the basis of laboratory experiments, Dainton (1954*a*, 1954*b*) postulated that slug activity is stimulated by falling temperature and inhibited by rising temperature, but is unaffected by light intensity after a period of habituation. However, Karlin (1961) and Getz (1963) have shown that, in the field, slugs are sensitive to changes in light intensity. Moreover, Henderson & Pelluet (1960) demonstrated that prolonged exposure to visible light produced degenerative changes in the ovotestis of *Agriolimax reticulatus* (Müller).

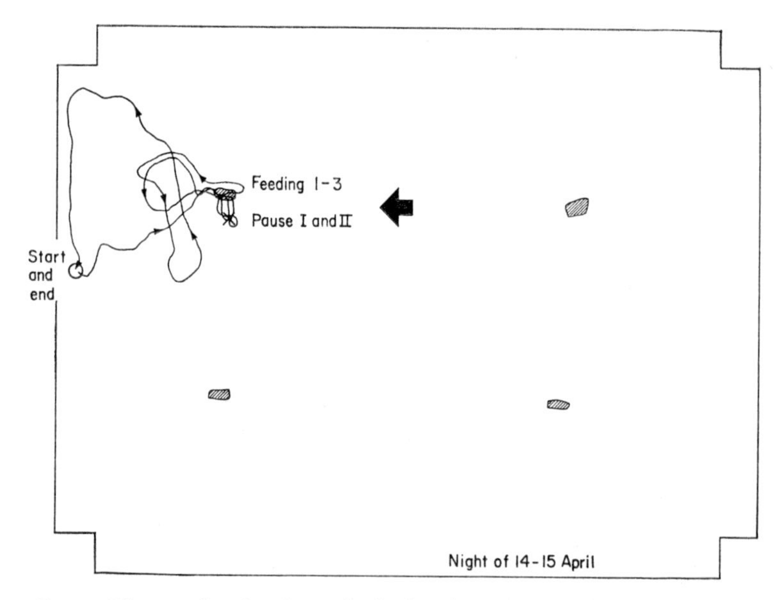

FIG. 1. Diagram showing the track of a slug during nocturnal surface activity. The 4 shaded rectangles show the positions of carrot slices used as food during filming. The boundary marks represent the walls of the arena and the arrow is a reference point. The arena is 80×100 cm.

If slugs respond to temperature changes, as believed by Dainton (1954*a*), then the soil temperature at the depth at which the slugs rest during the day must change sufficiently to serve as a cue for surface initiation of activity. South (1964) showed that most *A. reticulatus* in both arable and grassland sites rested in the top 10 cm of soil. This species normally becomes active on the surface between sunset and midnight and disappears below ground soon after sunrise. However, it cannot be assumed that all other species behave in a similar way, as it is known that some species crawl on the surface during daylight hours. When accurate measurements are made of the surface behaviour of slugs it is possible to subdivide this behaviour into a series of distinct occupations such as, crawling, feeding, resting or copulating (see Fig. 2).

In general the surface activity of soil animals is controlled by humidity, temperature and light intensity, and slugs react to all these factors. Slugs have well developed eyes and are known to be sensitive to temperature changes. Their skin is covered with mucus and they lose much water both by evaporation from their body surface and through the hydrated

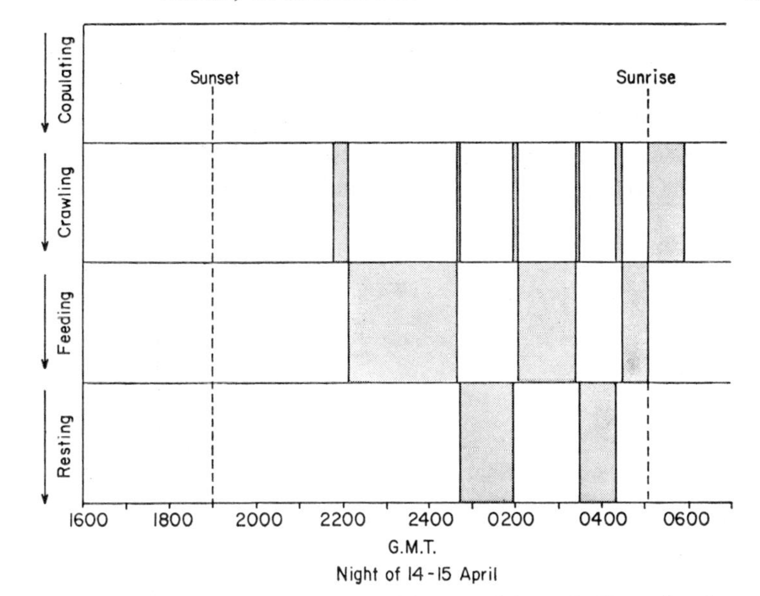

FIG. 2. Diagram taken from the track of the animal shown in Fig. 1, showing the durations of the various surface activities. The animal emerges at 21·45 hr and crawls, feeds, crawls, rests, crawls, feeds, crawls, rests, crawls, feeds and then crawls below the soil surface at 05·45 hr G.M.T.

mucus secretion that lubricates their path when crawling (Machin 1964). This paper is concerned with the measurement of the factors governing the emergence of slugs onto the soil surface and their subsequent disappearance below ground.

Because the surface crawling of *A. reticulatus* is normally limited to the hours of darkness, it seemed likely that the environmental cue, or cues, used by this species might be related to changes associated with nightfall and daybreak. Accordingly, records both of soil temperatures and light intensity were made in conjunction with a film record of surface activity. It was discovered that crawling was inhibited by photographic flood-lights but seemed unaffected by high-speed flash illumination, which was, therefore, used for making time-lapse cinematographic records.

Instruments used for measuring microclimatic features

Changes in soil temperatures and light intensity were recorded on an 8-channel Grant D-type recorder from a large wooden box (80 cm×

100 cm × 80 cm) partially filled with 45 cm of soil. This instrument was battery-operated. Temperature changes were measured by thermistors protected from damage by stainless steel cases. These probes seemed more reliable than those used by Long (1957). As slugs are normally active between 0 and 15°C, a recorder was chosen with a range of −5°C–20°C. The accuracy claimed by the manufacturers is ±0·25°C, or ±2 per

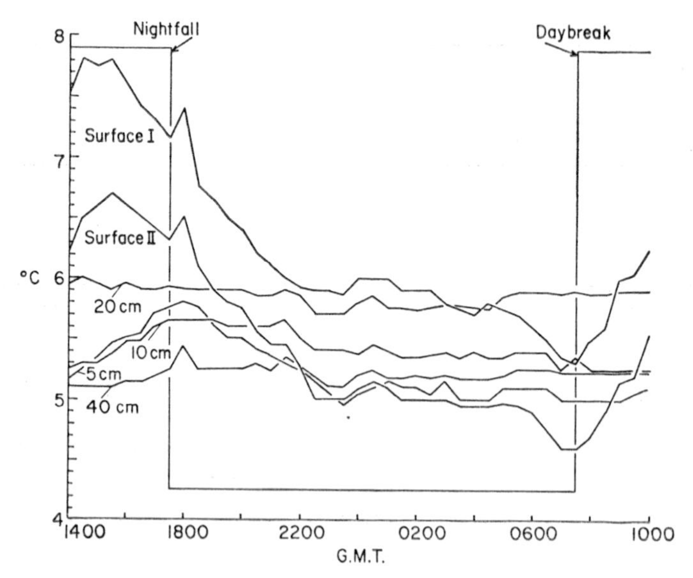

Fig. 3. Changes in soil temperatures on the night of 10–11 February 1967. Note the relatively small temperature difference between probes inserted at various depths. The data for this figure were taken from a trace obtained from a Grant recorder. Small temperature fluctuations have been eliminated by only using half-hourly readings.

cent of the full scale deflection. When this instrument was calibrated against a Griffin & George mercury-in-glass thermometer, calibrated by the National Physical Laboratory, it was found that under optimum conditions temperatures could be reliably measured to ±1 per cent of the full scale deflection. Temperature variations between probes could be read off the chart to 0·1°C. It seemed unlikely that slugs could respond to smaller temperature differences.

Economies in batteries and chart paper were effected by automatically switching on the recorder every 15 min. Since speed of response was not important, a robust, steel-cased, type of thermistor probe was used. A recording cycle of the 8 channels used 1·3 cm of chart paper, each roll lasting 16 days. A typical record of the changes of temperature in the soil over a 24-hr period is shown as a graph in Fig. 3.

Light intensity on the surface of the experimental arena was measured by means of a small photoelectric cell connected to one of the input channels of the recorder. Unlike that of the thermistors, the output from this cell was non-linear, and no calibration of its performance was made; the essential requirement was that it should record nightfall and daybreak.

Correlation of data

To synchronize the meteorological records with the behavioural (film) records, the start of temperature and light recordings was noted on the chart, which enabled the time of all subsequent records to be calculated. Behavioural records were made by time-lapse motion photography of slugs crawling on the surface of an arena partially filled with soil. The time spent feeding, crawling and resting was noted by simultaneously photographing a small clock placed within the arena.

Thus, environmental factors could be correlated with surface behaviour of slugs within the limits set by the 15-min recycling time of the Grant recorder. In practice it was found that night temperatures rarely fluctuated more than 0·25°C between recordings, and so temperatures at times between recordings could be reliably estimated.

Results

Measurements over successive 24-hr periods show that soil temperatures fell steadily during the evening and throughout the night, usually reaching a minimum at sunrise. Such changes do not appear to provide an adequate cue for the fairly rapid alterations in behaviour which are shown in Fig. 2 to be phased to coincide with major changes in light intensity associated with night and day.

Discussion

Dainton (1954*a*), using an actograph, showed that slugs under uniform conditions had a 24-hr rhythm of activity. The results presented in the present paper can be interpreted in the following way: activity in the soil is initiated by an inherent rhythm; when the slugs approach the surface, they are deterred from emerging if the eye tentacles encounter bright light; they do not emerge until after a low threshold value is reached, usually after sunset. However, the time of day that this threshold light value occurred varied greatly with the weather and the aspect of the arena. It was necessary to use a plant-free arena in which the light intensity at the surface was uniform. Although nocturnal crawling often finishes before sunrise, when slugs are on the surface, increased light intensity prior to sunrise increases the rate of crawling and, since these animals are photonegative, thus initiates the crawl 'home'. Diurnal light changes therefore reinforce the natural rhythm.

Slugs crawling on the surface during the day might fall into two categories; those that were returning 'home', and those whose surface crawling had been triggered by a decrease in light intensity, as would occur during a rain shower.

This work was supported by a grant from the Central Research Fund.

REFERENCES

BARNES H.F. & WEIL J.W. (1942) Baiting slugs using metaldehyde mixed with various substances. *Ann. appl. Biol.* **29**, 56–68.

BETT J.A. (1960) The breeding seasons of slugs in gardens. *Proc. zool. Soc. Lond.* **135**, 559–68.

DAINTON B.H. (1954*a*) The activity of slugs. 1. The induction of activity by changing temperatures. *J. exp. Biol.* **31**, 165–87.

DAINTON B.H. (1954*b*) The activity of slugs. 2. The effect of light and air currents. *J. exp. Biol.* **31**, 188–97.

GETZ L.L. (1963) Light preferences of molluscs. *Ecology* **44**, 612–13.

HENDERSON N.E. & PELLUET D. (1960) The effect of visible light on the ovotestis of the slug *Deroceras reticulatum* (Müller). *Can. J. Zool.* **38**, 173–8.

HOWITT A.J. (1961) Chemical control of slugs in orchard grass-Ladino white clover in the Pacific Northwest. *J. econ. Ent.* **54**, 778–81.

KARLIN E.J. (1961) Temperature and light as factors affecting the locomotor activity of slugs. *Nautilus* **74**, 125–30.

LONG I.F. (1957) Instruments for micro-meteorology. *Q. Jl R. met. Soc.* **83**, 202–14.

MACHIN J. (1964) The evaporation of water from *Helix aspersa*. I. The nature of the evaporating surface. *J. Exp. Biol.* **41**, 759–69.

SOUTH A. (1964) Estimation of slug populations. *Ann. appl. Biol.* **53**, 251–58.

WEBLEY D. (1964) Slug activity in relation to weather. *Ann. appl. Biol.* **53**, 407–14.

THE ASSESSMENT OF SOIL WATER STATUS AS IT AFFECTS PLANT WATER USE

E. J. WINTER
National Vegetable Research Station, Wellesbourne, Warwick

SUMMARY The effect of soil moisture environment on a plant community depends upon current soil moisture status, and on factors which influence abstraction of water by roots, including weather parameters, soil texture and moisture release characteristics, the extent and concentration of absorbing rootlets, and the rate of movement of water through the soil.

The term 'exploitable water' describes that proportion of the soil available water which can be taken up by particular plants under prevailing conditions.

Overall removal of water from the soil may be measured directly, as with weighable lysimeters, or inferred from determinations or estimates of the evaporative power of the atmosphere; the soil location from which water has been removed may be plotted by means of *in-situ* instruments such as tensiometers, resistance blocks or the neutron probe, or by the examination of abstracted soil samples. Suitable techniques are described.

The limitations of lysimeters are discussed, especially the difficulty of producing the same environment inside a container as prevails outside it.

In the following account soil water/plant relations are treated from the viewpoint of the biologist who is primarily interested in the plants; soil water status has been described in terms related to growing plants in preference to physical terms related to the soil itself.

INTRODUCTION

The effect of the soil water environment on a given plant community depends on the total soil moisture content and on factors which influence the abstraction of water by roots; the purpose of this paper is to describe these factors and some methods of measuring them. It is comparatively easy to determine the total water content of the soil; it is not so easy to decide how much of the total water content is available to plants or to specify the effects on water uptake and plant growth of a soil moisture status below the maximum possible.

Bulletin 23 of the International Soil Science Society (1963) gives authoritative definitions of terms relating to water in the soil, based on the concept that such water is subject to forces derived from the presence of solid particles, dissolved salts, gas pressure and gravity. For transfer of water to take place into the atmosphere by evaporation from the soil surface, or by transpiration from foliage, energy must be introduced into the system from outside. Thus the rate of water transport is governed by the absolute amount of water present, the magnitude of the forces holding it in the soil and the sources and quantity of energy available to overcome them. There are many methods for determining evaporation and transpiration based on estimates of the energy available in the environment and these will not be discussed in detail here (a new critical account of the methods is in course of preparation by the World Meteorological Organization). In general they yield values which may be regarded as maxima for the given climatic conditions when water supply is non-limiting; actual evaporation and transpiration rates are influenced by modifying factors, which include the soil type, condition and current water content, the kind of vegetation, and the extent and degree of ramification of its roots.

1 *Available water in soils*

Available water, sometimes defined as that part of the soil water which plants can extract sufficiently rapidly to maintain turgidity, may be assessed as the numerical difference between the soil water content at field capacity, the upper limit, and that present at permanent wilting percentage, the lower limit. The three terms, 'available water', 'field capacity' and 'permanent wilting percentage', are not included in the I.S.S.S. list referred to above because they cannot be precisely defined in physical terms, but they are nevertheless effectively and frequently used by biologists.

Permanent wilting percentage describes the moisture status of a soil below which plants cannot take up water sufficiently rapidly to maintain turgidity and growth even when placed in a saturated atmosphere; in physical terms, the matric suction of a soil at permanent wilting percentage is about 15 atmospheres. Field capacity is usually described as the status of a soil which is holding its maximum amount of water against free drainage, although slow drainage persists more or less indefinitely; in some soils after saturation there is a comparatively abrupt change from a fast to a slow drainage rate, usually within 48 hr, and the moisture status at the point of change can be regarded as field capacity. To obtain comparable results, and to include those soils which do not show this abupt change, it is usual to carry out field capacity determinations at the arbitrarily-set time of 48 hr after saturation. Some workers use a matric suction value of $\frac{1}{3}$ atmosphere as indicative of field capacity; others use a lower suction.

The imprecise definition of field capacity and the fact that its matric suction value can differ according to the time of year and tilth of the soil do not invalidate its use by biologists, but do make it essential to include with field capacity data a statement of the conditions under which the measurements were made.

Though field capacity is normally described as the upper limit of available water, sporadic rainfall may keep at least the surface layers of the soil above field capacity for appreciable periods. When this happens during the growing season, transpiration occurring before the soil has had time to drain down to field capacity takes place at the expense of the excess water, which should then be regarded logically as part of the available water. Wilcox (1962) has therefore proposed that the upper limit of available water should be defined as the highest moisture content of a soil which includes all moisture available for consumptive use but excludes all drainage below the root zone. Verigo & Razumova (1963) also recognize that plants can use water in soil above field capacity. Their definition of 'useful water' for their Russian conditions includes all the water from saturation to permanent wilting percentage. Long periods of frequent light rainfall, during which most of the transpiration demand is satisfied by soil water held above the field capacity level, are comparatively rare in U.K.; under average conditions the error introduced by regarding field capacity as the upper limit is unlikely to exceed a few tenths of an inch, an accuracy sufficient for most field purposes.

The description of soil water status in physical terms implies that not all the available water is equally available to plants, and that as the soil water is attenuated by evaporation and transpiration so the force holding the remaining moisture in the soil increases. This phenomenon is exploited in the tensiometer which measures directly the matric suction retaining water in the soil. The matric suction comprises that part of the total retaining force which is derived from the presence of the solid soil particles and the intervening pore spaces; the remainder of the total force, the osmotic suction, is derived from the presence of solutes and is not measured by the tensiometer or pressure membrane apparatus. Except for saline soils, where osmotic suction has been estimated routinely from the electrical conductivity of extracted water, few determinations of osmotic suction have been published. Recently, however, Williams (1966) measured matric suction, using pressure membrane apparatus, and then the osmotic suction of the same soil samples, using a thermocouple psychrometer. He concluded that for ordinary soils the osmotic suction was small enough to be disregarded.

2 *Exploitable water in soils*

The relationship between soil moisture content and matric suction retaining the water was defined by Childs (1940) as the soil moisture characteristic, and it is conveniently determined using pressure membrane apparatus. Attempts to establish a firm relationship between matric suction, plant water uptake and plant growth have met with only moderate success. Many workers have demonstrated that the establishment of water stress in the soil results in reduction in growth but it does not necessarily follow that, under these conditions, there is an accompanying reduction in water uptake and transpiration rate. The efficiency of water removal is influenced by the size of the root system, the concentration of absorbing rootlets and the rate of movement of water through the particular soil from unramified regions unexploited by roots towards those already occupied. This subject, the efficiency of exploitation of available water, has received little attention but is of practical importance in agriculture and in the study of natural vegetation and is worthy of further investigation.

3 *Depletion of available water*

In irrigation experiments watering treatments are often based on the permitted depletion of a specified percentage of the available water, but it is important to appreciate that depletion of the same percentage of the available water in two different soils does not necessarily impose the same stress on plants growing in them even though the definitions imply that the field capacity stress and the wilting percentage stress, respectively, are the same in all soils. Soils may differ in respect of the *shape* of the moisture release curve but not in its upper and lower limits. Basing experimental watering treatments on a specified depletion of available water enables findings to be applied to a soil of a different type from that on which they were obtained, only if the moisture release characteristic curve of each soil is known.

4 *Available water/texture relationship*

The available water capacity of a soil has been shown by Salter & Williams (1965) to be sufficiently closely related to its particle size composition, i.e. its texture, to enable the available water capacity of a soil to be predicted in the field from a knowledge of its textural class, with an accuracy of about ± 10 per cent.

5 *Capillary movement*

Upward capillary movement of water in soil is normally considered to be slow as evidenced by the formation of a dry crust on a soil which nevertheless retains a high moisture content below the surface. However, in recent low-speed wind tunnel experiments by the author, soil with a comparatively small crumb structure, in containers, dried to a depth of several inches within about a week, and different soils dried at different rates. The undoubted success of the capillary bench for automatic watering of pot plants (Winter & Cook 1965) provides further evidence for the rapid upward movement of considerable quantities of water under suitable conditions.

6 *The study of soil water removal by plants*

Overall water loss from soil may be measured directly, using weighable lysimeters, or inferred from determinations or estimates of the evaporative power of the environment. Such methods give no indication of the source of the water, i.e. from which part of the soil it has been lost. For this, moisture status must be assessed at known depths, using, for example, the neutron probe, tensiometers or other *in situ* sensors, or removing soil samples for analysis. The latter is frequently regarded as the most reliable method, but is laborious and has the serious disadvantage that it leaves auger holes which, even if refilled, can affect subsequent water movement in the soil. Any method which relies on discrete sampling is subject to error derived from the variability of field soil, which may be considerable; one advantage of the neutron probe is that, without disturbance other than the preliminary insertion of the sampling tube, it responds to the moisture status of the soil within a sphere of the order of 12 in. diameter; furthermore the moisture content of the identical soil may be repeatedly assessed with no damage or disturbance. For many purposes, interest lies in the integrated water status of a mass of soil of the same order of dimensions as the whole root zone, but if need be, the volume of soil to which the probe responds can be reduced by internal modifications, and satisfactory resolution between soil layers only a few inches deep has been obtained.

For a complete picture of the pattern of removal of soil water an overall method may be combined with a sampling method, e.g. a weighable lysimeter may have tensiometers inserted in it at various depths.

With the exception of tensiometry, all the methods mentioned yield results relating to total soil water content and for meaningful interpretation they must be used in conjunction with moisture release characteristic data of the particular soil under investigation. They may be used with simple available water data, i.e. field capacity and wilting percentage, but for reasons given above, this will not necessarily yield results comparable between different soils.

7 *Methods suitable for the field ecologlst*

The dynamic nature of the soil/water/plant system makes it essential for the ecologist to undertake repeated measurements on the soil in which he is interested; in the field this involves repeated sampling or the use of more or less permanently-installed sensors; an alternative is to remove soil cores for use as weighable lysimeters in which suitable plants may be grown under the comparatively controlled evaporating conditions of the laboratory glasshouse with or without moisture sensors at suitable depths. Undisturbed monolith cores (i.e. not containers filled with excavated soil) are needed for this purpose because of the virtual impossibility of reproducing the natural packing and arrangement of field soil layers in a filled lysimeter. The use of a core for studying changes in soil water status and plant growth is open to certain objections; firstly, unless a device such as a suction plate is inserted beneath the core, the soil moisture tension within it is not identical with that at the same depth in field soil; for most biological purposes the resulting error is likely to be negligible. Secondly, the root range of plants grown in a container is necessarily limited, and though this objection might be met by using containers larger than the normal field root range of the plants, in practice even with comparatively small herbaceous plants such a container would need to be several feet deep.

The use of a container introduces into the system an undesirable soil/container interface down which water may flow atypically; obviously the larger the container diameter the smaller is the proportion of interface area to container volume. The effect of the interface is likely to be greatest at the time of watering; therefore watering should be done by one of the methods designed to nullify the interface effect, or the water relations of the culture should be studied only between waterings.

Perhaps the most practicable approach to the ecological study of soil water in the field is to record the weather environment and hence the evaporative potential of the site and to use this in combination with the available water and moisture release characteristic data of the local soil to estimate the stress to which the plant community has been subject. Such estimates need to be checked against gravimetric analyses of soil samples made at intervals throughout the season.

Accurate estimation of evaporative potential by the well-known meteorological methods requires the separate recording of various parameters of the environment, but several different instruments have been devised for recording evaporation directly. While it is easy to obtain results of limited accuracy, most of these instruments are severely affected by the vagaries of the site and although they have much to commend them for ecological use, care must be taken in locating them especially when their records are to be used in comparing one place with another. Such evaporimeters have been combined with rain collectors so as to record not only deficit of rain below evaporation, but also excess rainfall, i.e. run-off or drainage, an item of obvious interest to ecologists and hydrologists.

This account of current thought on plant/soil/water relations is intended to show that, to assess the effect of soil moisture on a plant community, information is needed on the available water capacity of the soil and its moisture release characteristics, the plant type, rooting depth and concentration, and current meteorological data. Some methods of making these determinations are listed in the Appendix.

APPENDIX

METHODS FOR ASSESSING THE EFFECT OF SOIL WATER STATUS ON PLANTS

(Those procedures used by the author and his colleagues are marked thus ★. For information about other procedures the reader is referred to the original authors' publications.)

1 *Current soil water status*

This may be inferred from estimates or measurements of the evaporation which has occurred since the soil was last at a known water status, or it may be measured directly.

(i) *Meteorological estimates*

The many methods available have been critically reviewed by several authors (e.g. Deacon, Priestley & Swinbank 1958; Linacre 1963). For use with vegetable crops which cover the soil completely for only about 20 per cent of their field life, Penman's original method has been extended by including empirically determined factors to take into account the percentage area of bare soil, the season of the year, the current soil moisture deficit and the period since the soil surface was last wetted★ (Winter Salter & Stanhill 1959). The basic procedure has been programmed for computer operation★ (Berry 1964).

(ii) *Meteorological measurements*

Whereas meteorological estimates use averaged and integrated values for many parameters and so tend to give an overall picture of the evaporating conditions affecting a plant community, direct measurement of water loss at a single point may be influenced by local peculiarities. Subject to the limitation that an evaporimeter may supply appreciable quantities of water to the air immediately round it and so influence the neighbouring microclimate, the direct measurement of evaporation is usually simpler than the estimating procedure.

Stanhill (1962) has used Piche evaporimeter readings to substitute the aerodynamic term in Penman's equation, while Holmes & Robertson

(1958) have used an atmometer, consisting of a horizontal wetted alundum disc for directly scheduling commercial irrigation.

The U.K. 6 ft square sunken evaporation tank* (Lapworth, Glasspoole & Lloyd 1948) has been found to give a reasonable approximation to estimated evaporation from an open water surface. Many workers use the USDA Class A 1 m diam pan, but this is more complicated in installation and operation than the U.K. tank.

Hudson's simple dish evaporimeter (1963) has given consistent results under a wide variety of conditions. Winter (1963*a*) and Salter have developed a deficit/surplus indicator* in which a rain collector was combined with an evaporimeter so that the net reading corresponded to soil water status. In a recent model, the area of the evaporating surface can be automatically reduced as the deficit rises, thus simulating the reduced evaporation and transpiration rate expected as the soil moisture content falls.

The simple transpirometer of Garnier* (1952) has been widely used in U.K. It consisted of a drainage lysimeter in which the growing crop received a measured excess of water daily, and the difference between this and drainage was taken as transpiration loss; provided that due account was taken of water 'stored' in the contained soil above field capacity (Green 1959) reproducible and meaningful results were obtained.

(iii) *Gravimetric determination of the moisture content of soil samples**

Because of the variability of natural soil it is essential to replicate adequately and to mix soil taken from appropriate depths in several bores. Salter & Williams, in the work referred to above, dried samples (about 160 g) to constant weight ($\pm 0{\cdot}1$ g) for a minimum of $4\frac{1}{2}$ hr at 105°C. Apparent specific gravity, for converting results to a volume/volume basis, was measured by determining the dry weight of soil in a 6 in. diam steel tube filled by being driven into undisturbed soil.

(iv) In situ *methods for determining soil moisture status*

The neutron scattering method is the most convenient for rapid and repeated determinations at numerous sites, but the apparatus is complex and expensive. Van Bavel, Nixon & Hauser (1963) have discussed its development and current status. Cheaper *in situ* methods include the use of electrical capacitance (e.g., Lejeune & Arnould 1958), thermal conductivity (Lindner 1964), high-frequency electro-magnetic waves (Van der Westhuizen 1964) and photo-electric reflectivity (Jacob 1962).

The electrical resistance method (e.g., Bouget, Elrick & Tanner 1958) has been widely applied, using plaster of paris, nylon, fibreglass, or combinations of these for the absorbent medium. The units are sensitive to temperature and soluble salts, their response exhibits hysteresis and they must be individually calibrated to suit the particular soil. They are especially useful for indicating when the wetted front reaches them during precipitation.

Many different patterns of tensiometer have been developed for research and commercial use; the simple equipment described by Webster (1966)* could readily be adapted for field ecological use. Tensiometers can operate only to a suction of rather less than one atmosphere. In an average soil this might correspond to depletion of about half of the available water.

2 *Soil available water capacity*

This is usually expressed on a volume basis, and in inches or centimetres per foot or metre depth and is obtained by the formula:

$$AWC = \frac{(FC - PWP) \times ASG \times \text{depth}}{100}$$

where *FC* and *PWP* are percentage soil moisture content at field capacity and permanent wilting percentage respectively, and *ASG* is the apparent specific gravity.

(i) *Field capacity*

Salter & Williams (1965) have shown that the determination of field capacity by subjecting undisturbed, wetted soil samples to $\frac{1}{3}$ atmosphere in pressure membrane apparatus gave results often substantially lower than determinations made by thoroughly wetting field soil *in situ*, allowing to drain for 48 hr while preventing surface evaporation, and then measuring the moisture content of soil samples.*

(ii) *Permanent wilting percentage*

They also compared results obtained with disturbed and undisturbed soil samples using the sunflower technique (Salter & Haworth, 1961) and the 15 atm pressure membrane technique. They concluded that it was desirable to use undisturbed soil cores and that the two techniques gave similar results with most of the soils tried.*

(iii) *Moisture release characteristic curve* ('moisture characteristic')
Although undisturbed cores would be expected to yield results closest to the release properties to be found under natural conditions, many workers have for convenience used samples of air-dried, sieved soil. Salter & Williams (1965) determined characteristic curves in pressure membrane apparatus after treating the soil samples in various ways. They stressed that the availability of water to plants is expressed by the *shape* of the percentage moisture/log tension curve, and they found that sieving and air-drying materially affected the shape of this curve compared with that obtained using naturally moistened undisturbed cores, stored untrimmed in a water-saturated atmosphere until required.*

(iv) *Rapid field assessment of soil available water capacity*
In Salter & Williams' method (1967), soil samples obtained from each horizon were judged separately for textural class simply by rubbing the moistened soil between the fingers. Using a special pocket slide rule (Williams 1967) the available water capacity in relation to texture and thickness of each horizon was totalled to give an overall figure for the whole profile.*

(v) *Lysimetry*
In addition to the many kinds of lysimeters using more or less conventional lever weighing mechanisms,* weight changes in soil masses have been continuously recorded by means of strain gauges (Hand 1966) displacement (King, Tanner & Suomi 1956), air pressure (Berwick 1966), hydraulic pressure (Forsgate, Hosegood & McCulloch 1965), and many other ingenious systems. The multiplicity of designs indicates the evident need for weighable lysimeters and the dis-satisfaction of new workers with existing equipment. The precision of most lysimeters is roughly inversely related to their cost. To obtain, at reasonable cost, adequate replication for dealing with variable soil and plant material, Winter (1963*b*) devised a simple hydraulic lysimeter using readily available mass-produced components*; though of relatively low precision, this instrument would readily detect one day's water loss under field conditions.

ACKNOWLEDGEMENT

I wish gratefully to acknowledge the contribution made by my colleagues Dr Salter and Mr Williams who carried out much of the work reviewed in this paper.

REFERENCES

Berry G. (1964) The evaluation of Penman's natural evaporation formula by electronic computer. *Aust. J. Appl. Sci.* **15**, 61–4.

Berwick P.D. (1966) A pneumatic lysimeter. Divn. Met. Physics, CSIRO Aspendale Vic. Aust. Private communication.

Bouget S.J., Elrick D.E. & Tanner C.B. (1958) Electrical resistance units for moisture measurements: their moisture hysteresis, uniformity and sensitivity. *Soil Sci.* **86**, 298–304.

Childs E.C. (1940) The use of soil moisture characteristics in soil studies. *Soil Sci.* **50**, 239–52.

Deacon L.L., Priestley C.H.B. & Swinbank W.C. (1958) Evaporation and the water balance. *UNESCO Arid Zone Res.* 10, *Climatol., Rev. Res.* 9–34.

Forsgate J.A., Hosegood P.H. & McCulloch J.S.G. (1965) Design and installation of semi-enclosed hydraulic lysimeters. *Agric. Meteorol.* **2**, 43–52.

Garnier B.J. (1952) A simple apparatus for measuring potential evapo-transpiration. *Nature, Lond.* **170**, 286–7.

Green F.H. (1959) Some observations of potential evaporation. *Quart. J. Roy. met. Soc.* **85**, 152–8.

Hand D.W. (1966) N.I.A.E. Silsoe Bedford—private communication.

Holmes R.M. & Robertson G.W. (1958) Conversion of latent evaporation to potential evapotranspiration. *Can. J. Pl. Sci.* **38**, 164–72.

Hudson J.P. (1963) Variations in evaporation rates in Gezira cotton fields. *Emp. Cotton Growing Rev.* **40**, 253–61.

International Society of Soil Science (1963) Soil Physics Terminology. *Bull. Internat. Soc. Soil Sci.* **23**, 7–10.

Jacob H.P. (1962) A probe for the photoelectric measurement of soil moisture. *Z. Landeskultur* **3**, 328–32.

King K.M., Tanner C.B. & Suomi V.E. (1956) A floating lysimeter and its evaporation recorder. *Trans. Amer. Geophys. Union* **37**, 738–40.

Lapworth C.F., Glasspoole J. & Lloyd D. (1948) Report on standard methods of measurement of evaporation. *J. Inst. Wat. Engnrs* **2**, 257–66.

Lejeune G. & Arnould G. (1958) The determination of moisture in soils and solid substances by means of an apparatus based on the changes in the dielectric constants. *C. R. Acad. Sci. Paris* **246**, 1217–9.

Linacre E.T. (1963) Determining evapo-transpiration rates. *J. agric. Sci.* **29**, 165–77.

Lindner H. (1964) An instrument for measuring the water content and temperature of soil by means of thermistors. *Albrecht-Thaer-Arch.* **8**, 79–87.

Salter P.J. & Haworth F. (1961) The available water capacity of a sandy loam soil. I. A critical comparison of methods of determining the moisture content of soil at field capacity and at the permanent wilting percentage. *J. Soil Sci.* **12**, 326–34.

Salter P.J. & Williams J.B. (1965) The influence of texture on moisture characteristics of soils. II. Available water capacity and moisture release characteristics. *J. Soil Sci.* **16**, 310–7.

Salter P.J. & Williams J.B. (1967) The influence of texture on the moisture characteristics of soils. IV. A method of estimating the available water capacities of profiles in the field. *J. Soil Sci.* **18**, 174–81.

Stanhill G. (1962) The use of the Piche evaporimeter in the calculation of evaporation. *Quart. J. Roy. met. Soc.* **88**, 80–2.

Van Bavel C.H.M., Nixon P.R. & Hauser V.L. (1963) Soil moisture measurement with the neutron method. *U.S. Dep. Agri. Res. Serv.* 41/70, pp. 39.

Van der Westhuizen M. (1964) On the possibility of measuring soil moisture with high frequency electro-magnetic waves. *S. afr. J. agric. Sci.* **7**, 589–90.

Verigo S.A. & Rasumova L.A. (1963) Soil moisture and its significance in agriculture. Leningrad. Russian text. (English translation by National Lending Library for Science and Technology, Boston Spa.)

Webster R. (1966) The measurement of soil water tension in the field. *New Phytol.* **65**, 249–58.

Wilcox J.C. (1962) Rate of soil drainage following irrigation. III. A new concept of the upper limit of available water. *Can. J. Soil Sci.* **42**, 122–8.

Williams J.B. (1966) Studies on the measurement of soil moisture stress. M.Sc. Thesis Sheffield University.

Williams J.B. (1967) A device for calculating available-water. *Expl. Agric.* **3**, 159–62.

Winter E.J. (1963*a*) A valveless soil moisture deficit indicator. *J. agric. Engng Res.* **8**, 252–5.

Winter E.J. (1963*b*) A new type of lysimeter. *J. hort. Sci.* **38**, 160–8.

Winter E.J. & Cook R.D. (1965) Automatic watering in glasshouses. *Grower* **63**, 1290–1.

Winter E.J., Salter P.J. & Stanhill G. (1959) Lysimetry at the National Vegetable Research Station, Wellesbourne, Warwick, England. *Int. Assoc. Sci. Hydrol.* Pub. 49, 44–53.

THE QUANTITATIVE DESCRIPTION OF SOIL MOISTURE STATES IN NATURAL HABITATS WITH SPECIAL REFERENCE TO MOIST SOILS

V. I. Stewart & W. A. Adams
University College of Wales, Aberystwyth

SUMMARY Comparison of soil moisture states is often made on the basis of single parameters. These comparisons are seldom successful, either because no account is taken of the relevance of the occasion chosen for sampling, or because the results are expressed in a manner which has no immediate biological significance. The choice of sampling technique and of the method of expressing the results is reviewed. It is suggested that, in regions with a wet season, the minimum data reported should describe for each horizon of the soil profile the extent to which the pore space is full of water on a field capacity day. These values will represent the balance between air and water in the soil, a feature which is of supreme importance to the organisms present.

Soil water has a dominating influence on the pattern of soil profile development. It is a primary weathering agent affecting both chemical and physical weathering. Through its selective influence on the organisms present it may have a marked effect on the incorporation of organic matter. Through its influence on biological activity it may control the rate of organic and mineral decay and the nature of the end products produced. It is itself a medium for both passive and active transport of the products of mineral weathering and organic decay. It is largely responsible, therefore, for the pattern of redistribution of soil constituents which are expressed in the many distinctions of soil colour and form that are used by the pedologist to describe, interpret and classify soils.

Pedologists use gley characters related, principally, to the form and distribution of iron within the profile as evidence of drainage state but seem never to support their assumptions by any measured, quantitative evidence describing actual moisture states of any significance for drainage. However, the evidence in contributions to the *Journal of Ecology* since 1960 suggests that pedologists are not alone in being reluctant to get to grips with this problem. For this reason we think it may be worthwhile to take up the challenge and see, for example, what general advice we would give to soil and vegetation surveyors, or contributors to the *Biological Flora* who may wish to include useful soil moisture measurements in accounts of their work.

Although we are aware of the fact that there are many sophisticated ways of approaching this task, and much instrumentation to choose from, we will keep in mind the general aim of maximum reward for minimum effort.

There is, of course, good reason for the general lack of useful data on field moisture states. Practical problems associated with any attempt to give numerical expression to a soil moisture state are many, and most of these become acute in ecological studies in natural habitats. This is especially so where the soils are podzolic and it becomes necessary to compare the wetness of horizons which, within the same profile, may range from pure organic matter in layers representing various stages of decay, to mineral layers, often stony, and varying widely in texture and structural stability. These are not the best conditions for the placement of instruments or the taking of intact samples. However, this paper is not concerned with techniques as such but rather with modes of expressing soil moisture states.

The problem is to express differences in degrees of soil wetness on scales that are easy to interpret. Consideration will be given to: (1) the importance of choosing a significant occasion on which to sample; (2) the place where the measurement should be made and the type of sample to be taken, and (3) the concepts of wetness which are relevant and the type of measurement which is therefore required. So often numbers quoted relate to a concept of wetness which has no relevance, or which, standing on its own, cannot be interpreted.

Ultimately our confidence in the relevance of results obtained by the use of any particular technique intended to provide information on field states, must pass the test of credibility when applied to known situations. For example, a technique for the estimation of soil moisture which consistently failed to show a tendency towards increasing wetness for soils in hollows compared with those on slopes would not inspire confidence. Such is the form of test with which we shall be mainly concerned in this paper.

The procedure used to obtain the data presented for discussion involved direct sampling, into special containers, of intact cores. Points on the moisture characteristic of these intact samples were determined in the laboratory on a soil centrifuge supplied by Measuring and Scientific Equipment Limited, London, England. The general principles involved in the centrifuge procedure have been described by Piper (1950), and have been criticized recently by Wedler (1965). We would not necessarily commend this method for general use but to us, when used for intact field cores, it produces repeatable and credible results, and allows the possibility of partial calibration of intact cores directly from the field without any need for the structurally disruptive effect of prior saturation. (Details to be published later).

Choice of sampling occasion

The amount of water present in a soil depends upon the balance between such factors as: rainfall, interception, evaporation, transpiration, surface run-off and drainage. Its distribution within the soil depends upon site factors related to position on the slope, the micro-structure of the soil profile affecting moisture retention, and the impact of recent weather conditions modified through the local influence of vegetation. It is unlikely, therefore, that the soil moisture characteristics of a habitat will be adequately described by one figure, or even by a series of figures derived from a number of different measurements made on one occasion, unless the occasion is well chosen for the purpose.

Choice of occasion can be crucial in obtaining evidence of site drainage to relate to profile morphology. Waiting for such an occasion in soil moisture studies normally amounts to waiting for suitable weather conditions so that the variation in response of different soils to a given input of water can be assessed. This means waiting for a significant natural event, which may be tedious for those in a hurry but for those with patience, may reduce the actual work involved compared with that required for any random or objectively systematic method of sampling.

In humid climates, or areas with a wet season, a particularly significant occasion for the examination of natural soil drainage patterns is the third day following a period of heavy rain during, or immediately following, a generally wet period. This we would identify as 'a wet season, field capacity day'. Freely drained soils will then be close to field capacity, less well drained soils will still be receiving ground water from the drainage of profiles further up the slope, and horizons affected by local impedance within the profile may still retain excess water received from above.

Choice of sampling depth

For general descriptions of natural habitats there would seem to be every reason for sampling with reference to described horizons when these are clearly defined. The trouble in some natural soils is the large number of horizons that may warrant distinction and the fact that some of these may be quite thin. It may not be reasonable to sample features like thin iron pans, or drainage channels around prismatic structures, but the position of samples with reference to such features should certainly be recorded.

Taking the sample

How to sample is largely a matter determined by the technique of measurement but, when direct sampling is used, great care must be taken to keep the structure of the soil intact. Errors due to sample disturbance may not be crucial in strongly granular plough layers but cannot be ignored in the relatively unstable compact structures of podzol profiles.

Choice of basis for expression

Once a decision has been made on the time, place and manner of taking samples there still remains the question of what to measure and how to express the result.

Although a figure describing the amount of water passing laterally or vertically through different horizons within soil profiles might well provide information of considerable interest, the amount of fieldwork and instrumentation involved in studies of this kind puts considerations of parameters of rate outside the scope of what is to be considered in this paper. We shall consider mainly parameters of concentration and capacity.

A concentration may describe either the proportion of the whole that is water or the ratio between water and another component, for example, organic matter or total pore space. The figures themselves may be either weights or volumes.

The figures in Table 1 represent various concentration values which have been used to describe moisture states in soil. In this instance the situation being described is the relative wetness of different horizons in three soils which are related by their position on the slope and which

TABLE 1. Comparison of various moisture concentration parameters used to describe the wetness of a slope sequence of soils in a permanent pasture field on the College Farm, U.C.W., Aberystwyth. Profiles sampled by intact cores at fixed depths, within defined horizons, on Field Capacity Day, 14 May 1959.

Soil Series Position on Slope Drainage Class	Denbigh Upper slope Free	Sannan Break in slope Imperfect	Cegin Hollow Poor	Sampling Depth in. (cm)
Horizon	*A*1	*A*11	*A*1*g*	
Per cent fresh wt. total sample	38	34	40	
Per cent fresh wt. fine earth	38	34	40	
Per cent oven dry wt. fine earth	62	52	67	3–6
Per cent field vol. total sample	50	47	55	(7–15)
Per cent field vol. fine earth	50	47	55	
Per cent oven dry vol. fine earth	70	55	81	
Per cent total pore space field sample	71	77	84	
Coefficient of humidity	8·8	4·6	7·8	
Horizon	(*B*)1	*A*12	G1	
Per cent fresh wt. total sample	17	22	33	
Per cent fresh wt. fine earth	24	27	33	
Per cent oven dry wt. fine earth	31	37	49	12–15
Per cent field vol. total sample	24	26	46	(30–38)
Per cent field vol. fine earth	25	29	46	
Per cent oven dry vol. fine earth	35	38	54	
Per cent total pore space field sample	42	40	71	
Coefficient of humidity	9·3	6·9	16·3	
Horizon	(*B*)2	(*B*)*g*	G2	
Per cent fresh wt. total sample	12	27	21	
Per cent fresh wt. fine earth	21	32	22	
Per cent oven dry wt. fine earth	26	47	29	21–24
Per cent field vol. total sample	19	36	39	(53–61)
Per cent field vol. fine earth	25	39	40	
Per cent oven dry vol. fine earth	32	49	36	
Per cent total pore space field sample	36	57	86	
Coefficient of humidity	14·4	13·4	36·2	

occur together in a permanent pasture field on the College Farm at Aberystwyth. They were sampled in spring on the third dry day after a period of fairly heavy rain—1·02 in. (27 mm) in the previous fortnight, 3·5 in. (89 mm) in the previous month.

Anyone digging these profiles on the sampling day would have been impressed by the obvious association of wetness and gley characters. The surface of the Cegin profile was clearly mottled and the sub-surface horizons were strongly gleyed. The Sannan profile was also strongly gleyed below 2 ft but, the lowest horizon from this profile, represented in Table 1, showed only slight signs of gleying in the form of occasional manganese nodules.

Any useful parameter of wetness should reflect these obvious trends in gleying as they are typical of what might be expected in a normal slope sequence, but of all the numbers quoted in Table 1, only the seventh parameter which relates water content to total pore space actually does so.

Soil is a variable mixture of organic matter, mineral matter, water and air, that is, it is made up of a variety of components differing widely in density. Furthermore, these components may be organized in a variety of ways leading to the creation of different structural forms, some resulting in far more pore space than others. As a result no weight can define a standard volume of soil and no volume, which involves the whole soil, can define a standard amount of pore space for the retention of water. It is not surprising, therefore, that values of concentration involving the whole soil are of little use on their own.

Eliminating stones from the comparison may be justified as stones can represent a very large weight or volume by comparison with their insignificant water holding properties and, in small samples, they may be responsible for a considerable amount of random variation. However, though their elimination from the water/soil ratio may improve replication, water content, expressed as a percentage of the fine earth, still does not give a reliable picture of the significance of differences between samples in the data quoted in Table 1.

Percentage pore space filled with water is the only parameter of wetness which standing alone, unsupported by other data, is clear in its message about soil wetness. Its strength lies in the fact that it relates two elements that are uniquely interdependent, representing the balance between air and water in the space available to either. Moreover, the state described is of supreme importance to the organisms present and bears directly on the circumstances likely to condition the appearance of gley characters.

Soil is a porous medium made up of a complex mixture of pore sizes.

When soil drains, therefore, it does so in such a way that the larger pores empty first whilst smaller pores remain full. Thus a soil which is half full of water is not a soil with all its pores half full of water but a medium which is a mosaic of aerobic and anaerobic environments. For this reason we would suggest that the term used to describe the extent to which the pore space is full of water in the field should be 'per cent waterlogged', or 'waterlogging percentage'.

Values for per cent waterlogged shown in Table 2 suggest that 70 per cent waterlogged on a field capacity day may be indicative of conditions conducive to gleying. However, we should like to know how far this condition is solely a product of site drainage or owes something to the water-holding properties of the horizons themselves, i.e. their inherent retentivity.

pF, or other units of soil water potential have sometimes been used in the *Journal of Ecology* to define soil moisture states in field studies. On its own a pF value merely indicates the maximum pore size holding water; it says nothing about the volume or proportion of the total pore space which the waterholding pores represent. For example, though a tension value below that equivalent to pF2 indicates the presence of water in pores larger than might be expected to be filled in freely drained horizons, this does not necessarily mean that the horizon is more waterlogged than one at pF2·7. A soil could be completely waterlogged at pF2·7.

However, a very modest attempt to assess one point on the moisture release curve of a soil sample can be of considerable help in distinguishing soil horizons which are wet because of their retentiveness rather than their position on the slope. We would suggest that, where it is possible to take intact field cores, these should be used to determine field moisture content, moisture retention at pF2, and apparent and true density for the calculation of total pore space. Calibration of the soil core at pF2, (100 cm tension), is recommended as it may be determined on a simple tension table (Clement 1966), and, from our figures, would appear to be close to the tension achieved by freely drained soils in the field on a wet season, field capacity day. Information of this type is given in Table 2 and plotted in Fig. 1.

On a graph relating per cent waterlogged in the field to per cent waterlogged at pF2 a theoretical line of slope 1 : 1 can be drawn indicating the expected plot of freely drained horizons. This line should be checked against the plot of horizons known to be freely drained in the field. Horizons holding water greatly in excess of that which would be predicted from a knowledge of their retentivity, will be either poorly

TABLE 2. Further comparison of soil moisture parameters used to describe soil profiles in a permanent pasture field, on the College Farm, U.C.W., Aberystwyth.

No.	Soil Series	Profile drainage	Horizon	Moisture as per cent oven dry fine earth on F.C.D. (14 May 1959)	Per cent waterlogged on F.C.D.	Per cent waterlogged at pF2	Water content at 2 ft (61 cm) in in. (mm): Wilting point	pF 2·52 (1/3 Atm)	pF 2·00 (1/10 Atm)	Total pore space	Available water pF 2·52—W.P.	Available water pF 2·00—W.P.	Available water Salter, Berry & Williams (1966)[3]	As previous column but including correction for stones
1	Denbigh	Free	A1	62	71	72	2·1	6·0	6·6	12·6	3·9	4·5	3·9	3·5
2			(B)1	31	42	47	(53)	(152)	(168)	(320)	(99)	(114)	(99)	(89)
3			(B)2	26	36	38								
4	Sannan	Imperfect	A11	52	77	82	3·1	7·9	9·0	14·8	4·8	5·9	5·2	4·7
5			A12	37	40	44	(79)	(201)	(229)	(376)	(112)	(150)	(132)	(119)
6			[1](B)g	47	57	61								
7	Cegin	Poor	A1g	67	84	85	4·1	9·5	10·6	13·9	5·4	6·5	4·8	4·6
8			[2]G1	49	71	66	(104)	(241)	(269)	(353)	(137)	(165)	(122)	(117)
9			G2	29	86	81								

[1] g horizon with slight mottling
[2] G horizon with gleying dominant
[3] AWC (in./ft) $= 1{\cdot}50 - 0{\cdot}0120$ (per cent coarse sand) $+ 0{\cdot}0123$ (per cent Int. fine sand) $+ 0{\cdot}302$ (per cent org. C.)

drained due to position on the slope or poorly drained because of impeding horizons within the profile. Horizons liable to be gleyed, due solely to their retentivity, will appear close to the line for freely drained horizons but at a point high on the scale for retentivity. Points far below the theoretical line will indicate horizons which probably have been prevented from attaining field capacity by being cut off from the full influence of surface or ground water, for example, by a thin iron pan.

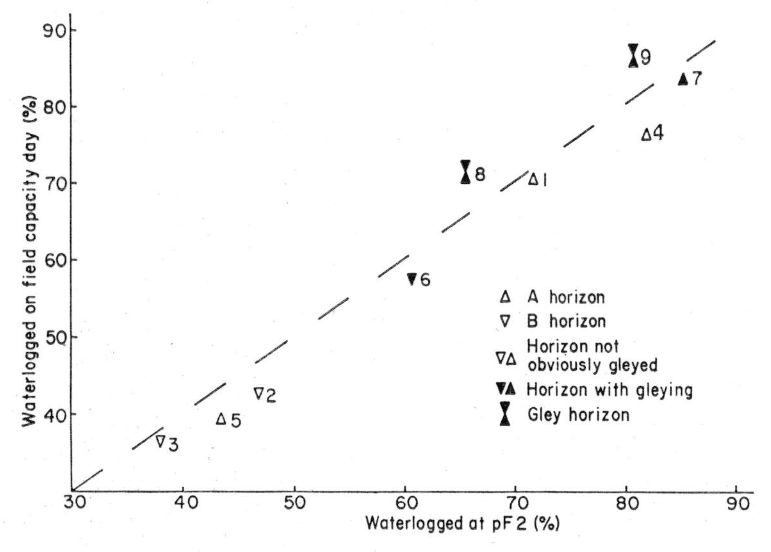

FIG. 1. Extent of waterlogging in horizons of soil profiles sampled on Field Capacity Day, 14 May 1959, and compared with the extent of waterlogging in the same samples at pF2. Samples taken from permanent pasture field, College Farm, U.C.W., Aberystwyth. Plot of data from Table 2.

A precise knowledge of the shape of the moisture release curve in the region of field capacity might be required in critical studies if the full significance of all deviations from the theoretical line for freely drained conditions were to be understood but a review of the data in the manner suggested above should allow major features affecting soil drainage to be distinguished.

The fully gleyed horizons in Cegin (Fig. 1) are clearly both highly retentive and poorly drained. The surface horizons of all the profiles, though not holding excess drainage water, are sufficiently retentive to suggest the possibility of gleying should these conditions persist for very long. It is of interest to note, therefore, that the organic matter content of all these surface horizons is relatively high (7–11 per cent) and all are liable on occasions to show signs of rusty mottling around roots although well developed root motting is a permanent feature of the Cegin surface only. The gley subsoil horizon in the Sannan profile occurs below 24 in. (61 cm) and is not included in the data used for the construction of Fig. 1. The lowest horizon in the Sannan profile which is included, the Bg horizon, is only sparsely mottled by manganese concentrations. Although this horizon is flushed by ground water under exceptional circumstances (see Table 3), it is apparently not excessively waterlogged on a normal, wet season, field capacity day.

When vegetation response is more likely to be related to the factor of moisture retention than the pattern of site drainage, as may be the case in dry areas, then consideration may have to be given to the possibility that neither field capacity nor retentivity, as defined, may be sufficiently precise to interpret vegetation response. However, it is unlikely that any static storage capacity can completely define the moisture supplying characteristics of a soil for this will depend very much on features of the vegetation, meteorological conditions affecting the rate of demand, the volume of the soil exploited by roots and factors in the soil affecting rate of replenishment (Cowan 1965). The data in Table 3 indicate that the pattern of soil desiccation during a dry season is related in West Wales to soil depth and drainage class and may be as well predicted by reference to field capacity or the moisture retained at pF2 as any precise estimate of available water (see Table 2). In the circumstances the indirect procedure for estimating available water proposed by Salter, Berry & Williams (1966) may be all that is justified but caution should be used in applying their formula outside the range of average agricultural soils where an abundance of stones, structural variability and high levels of organic matter are liable to occur.* For example, using their formula, an AO horizon, high in organic matter, would appear to retain more available water than it has volume to store.

General conclusions

When considerations of time and effort place a restriction on the amount of soil moisture data that can be recorded we recommend: (1) restricting sampling to significant occasions such as field capacity days; (2) making measurements which provide numerical evidence of waterlogging and which enable a distinction to be made between horizons poorly drained

* These limitations have now been acknowledged by Salter & Williams (1967).

TABLE 3. Record of progressive desiccation and subsequent wetting of soil profiles in permanent pasture field, College Farm, U.C.W., Aberystwyth, May to October 1959.

Soil series Position on Slope Drainage Class	Powys Crest Excessive	Denbigh Upper slope Free	Sannan Break in slope Imperfect	Cegin Hollow Poor	Sampling Depth in. (cm)
		pF Values			
14 May	1·9	2·1	2·4	2·1	
20 May	—	3·3	2·8	2·7	
9 September	4·5	3·9	3·7	3·8	3–6
5 October	5·1	4·3	4·0	3·9	(7–15)
20 October	3·6	2·6	3·1	3·2	
14 May		2·5	2·3	1·5	
20 May		2·5	2·4	2·0	
9 September	Rock	3·5	3·2	3·0	12–15
5 October		3·9	3·3	3·2	(30–38)
20 October		2·6	3·2	3·0	
14 May		2·3	2·3	1·3	
20 May		2·0	2·2	1·8	
9 September	Rock	3·2	2·9	3·1	21–24
5 October		3·6	3·1	3·4	(53–61)
20 October		2·9	0·7	2·2	

Water content to 2 ft (61 cm)

	Powys in.	Powys mm	Denbigh in.	Denbigh mm	Sannan in.	Sannan mm	Cegin in.	Cegin mm
14 May	3·8	97	6·3	160	8·4	213	11·0	279
20 May	—	—	4·9	124	8·1	206	11·2	284
9 September	0·9	23	3·6	91	5·9	150	6·6	168
5 October	0·7	18	3·1	79	5·2	132	5·5	140
20 October	1·3	33	5·1	130	8·0	203	8·7	221

Sampling conditions

14 May	Third dry day after period of moderately heavy rain.
20 May	Tenth dry day.
9 September	No rain after thunderstorm on 20 August. Grass scorched on Powys.
5 October	Virtually no rain for 46 days. Only Powys scorched.
20 October	Third dry day following end of drought. Heavy rain between 10 and 17 October.

because of site, and horizons liable to be waterlogged because of their retentivity.

In our view, these are modest but minimal requirements and, if accepted as such, would greatly enhance the usefulness of soil moisture data in descriptions of ecological situations. Had we quantitative evidence of this kind for all mapped soil series we should have useful numerical evidence of degrees of variation in soil drainage and an indication of the precise significance of gley characters in profile morphology.

Summary of practical procedure

1 Take samples on 'field capacity days' including samples from a standard freely drained soil for reference.
2 Take intact cores of known volume and weigh fresh.
3 Equilibrate field cores on tension table at pF2 and weigh.
4 Oven dry equilibrated cores and weigh.
5 Determine true density of each complete oven dry sample by displacement of carbon tetrachloride.
6 Determine total pore space in field core by applying the following formula:

$$\frac{\text{True density} - \text{Apparent density}}{\text{True density}} \times 100 = \text{Total pore space}/100\ \text{cm}^3\ \text{soil}$$

7 Express both (a) field moisture content and (b) moisture content at pF2 as percentages of the total pore space. These values indicate the degree of waterlogging in the moisture states examined.
8 If intact cores cannot be extracted for use in the laboratory then determine the volume of the hole from which the disturbed sample has been removed and report 7 (a) only.

ACKNOWLEDGEMENTS

The authors wish to acknowledge the technical assistance of Mr D.C.Pugh.

REFERENCES

Clement C.R. (1966) A simple and reliable tension table. *J. Soil Sci.* **17**, 133–5.
Cowan I.R. (1965) Transport of water in the soil-plant-atmosphere system. *J. appl. Ecol.* **2**, 221–41.
Piper C.S. (1950) *Soil and Plant Analysis.* University of Adelaide, Adelaide.

Salter P.J., Berry G. & Williams J.B. (1966) The influence of texture on the moisture characteristics of soils. III. Quantitative relationships between particle size, composition, and available-water capacity. *J. Soil Sci.* **17**, 93–8.

Salter P.J. & Williams J.B. (1967) The influence of texture on the moisture characteristics of soils. IV. A method of estimating the available water capacities of profiles in the field. *J. Soil Sci.* **18**, 174–81.

Wedler W. (1965) A method of pF value measurement using a centrifuge. *Z. Pfl.-Ernähr. Düng. Bodenk.* **109**, 249–60.

THE SIGNIFICANCE OF SOIL WATER TO SOIL INVERTEBRATES

D. R. GIFFORD
Department of Forestry and Natural Resources, University of Edinburgh

INTRODUCTION

Water is available to animals in the soil either as a fluid or as vapour. In both conditions the relationship between water and the animal is dynamic, a product of the stress between environmental water and the water content of the animal itself, modified by the physical properties of the animal's integument. Increment and disposal of water by other means than through the integument influence water content without dominating it in soil animals, which differ in this respect from most of the fauna living above ground. Soil arthropods are often sensitive to reduction in free soil water before atmospheric humidity in soil cavities is affected, but their water relations have only been studied from the latter viewpoint.

Water as vapour

Andrewartha & Birch (1954), Edney (1957) and Machin (1964*a*, *b*) have all considered evaporation formulae in a biological context. Their conclusions are based on the work of Ramsay (1935*a*) and Leighly (1937), and the expression given by Machin (1964*b*) takes into account the difficulties outlined by previous authors. Machin studied the evaporation from an extended snail, *Helix aspersa*, and considered the difficulties imposed by the geometry of the snail itself. Machin's expression is

$$E = K(p_o - p_d)\, c(v/x)^n,$$

where E is the evaporation rate, p_o and p_d are the vapour pressures of the evaporating surface and free air respectively, v is the wind speed, x the length of the evaporating surface parallel to wind direction. The terms c and n are empirically determined for each surface. K is a constant derived from $K = k\sqrt{\rho/\mu}$ where k is the coefficient of diffusion of water vapour, ρ is the air density and μ the molecular viscosity of air, all being temperature dependant.

Machin's expression obviates the need for direct consideration of turbulence and the geometry of the boundary layer by the introduction of his empirical terms. It is clear from his measurements of the boundary layer effect that it is deeper than the height of most soil animals placed on a plane surface in an airflow below 25 cm/sec; such wind velocities are applicable to the environment of animals in the litter stratum of the soil, but the evaporation stress on these is more complex even than Machin's model, since the turbulence and differential diffusion rates from surfaces of variable quality and aspect buffer water loss into the boundary layer. In such a system, where the tendency at all times is towards saturation vapour pressure, the expression given by Andrewartha & Birch (1954) is more easily applied

$$E = k \, . \, dp/dx$$

where k is a term similar to Machin's K, dependant on the physical effect of changing temperature on water vapour, and dp/dx describes the relationship between vapour pressure and the boundary layer x.

Lees (1946, 1947) demonstrated the ability of the sheep tick *Ixodes ricinus* to extract water from humidities as low as 86 per cent R.H. He suggested that this was achieved by active secretion inward from the hypodermal cells, which then made good the deficit by sorption through the cuticle. This capacity has not been demonstrated in soil animals, but Knülle (1965) studied a similar effect in *Acarus siro*, the grain mite.

Water as a liquid

Osmotic effects between soil water and animals living in continuous contact with it are desiccant only in special circumstances, such as in strongly saline conditions. Usually the osmotic pressure of body fluids is greater than that of the soil water, and the mechanism for regulation of body water has been extensively studied for the earthworm, the work being reviewed by Laverack (1963). The response of soil Protozoa to periodic water shortage has been reviewed by Stout & Heal (1967). Encystment is the usual result, but this may be brought about by high salinity or an inadequate respiratory regime, while excystment may depend on food availability, as well as the renewed presence of sufficient water.

The availability of water to arthropods is complicated by their common need to conserve that which they already possess. Water plays a fundamental role in their respiratory activity besides the obvious need for it in nutrition. Arthropod water relations are admirably reviewed by Edney (1957). The cuticle has been shown to be of two types, one wettable, and thus rather permeable to water and gases, the other much more extensive, and relatively impermeable, made so by a thin layer of grease or wax secreted on the epicutide. It is this film which allows the animal to conserve water. Our knowledge of the cuticle and its permeability is founded on experiments with free-living insects. Ramsay's demonstration (1935*b*) that with increasing temperature there was a sudden rapid increase in evaporation from cockroaches was followed by Wigglesworth's (1945) experiments on a number of insects for which he demonstrated critical temperatures where an abrupt increase in evaporation took place. Beament (1945) obtained the lipid materials from the cuticles of these insects and showed that they had melting points a little above the critical temperatures demonstrated by Wigglesworth. Holdgate (1955) and Noble-Nesbitt (1963*a*) investigated the wettability of cuticle, in particular the influence of surface geometry.

The reviews of Edney (1957), Beament (1961) and Wigglesworth (1965) provide an authoritative account of the recent work, but little of this has been concerned with true soil arthropods. These are small and their water relations may be confused by problems associated with cuticular respiration. Davies (1928) examined the sensitivity of some Colembola to desiccation and found that the only species with tracheate respiration examined, *Sminthurus viridis*, was much less vulnerable than other species. *Podura* is not a soil animal but Noble-Nesbitt's (1963*a*, *b*) study of its cuticle demonstrated that the wax layer was restricted to the tops of cuticular roughnesses, so that much of the cuticle was permeable.

Madge (1964*a*, *b*, *c*) studied the water relations of some soil Acari. In a study of *Belba geniculosa* he obtained critical temperatures of about 40°C for all the developmental stages as well as the adults, demonstrating the presence of a uniform waterproofing mechanism despite large changes in sclerotization of the cuticle during development. This was supported by a study of the tracheation; Madge ensured complete opening of the spiracles by increasing carbon dioxide concentration, but found no increase in transpiration occurred, in contrast to the results of Mellanby (1934) and Lees (1946) with animals of different habitat. Similar experiments with other Acari produced widely divergent results, some mites, such as *Hypochthonius rufulus* appearing to have a permeable cuticle, while other had critical temperatures as high as 50°C. Despite these results Madge found that *B. geniculosa* selected high humidities when given a choice, and had well developed humidity sensors on the forelegs.

The response of small arthropods to temperature and humidity gradients has been studied largely to check the efficiency of extraction apparatus. The work of Macfadyen (1953, 1961, 1962), Nef (1962), Kempson, Lloyd & Ghelardi (1963) and Block (1966) may be cited as examples. The marked succession in the reaction to gradients suggests that many animals are prompted to migrate before a saturation deficit occurs. Machin's (1964*a*) results suggest that, while there is free water in the soil, cavities will have a water saturated atmosphere unless very large, i.e. over 5 mm for the minimum dimension, when isolated from rapidly moving air (<15 cm/sec). Gifford (unpublished) set up a windtunnel to test the separate effect of some of the components of evaporation at the surface of forest litter. By varying temperature, humidity and windspeed independently, regressions of these variables on the mean depths in the litter of the microarthropods were determined. The most significant variable measured was water loss, with only scattered direct effects from the independent variables. If it is correct to accept the concept of atmospheric saturation to a late stage in the drying process, such movements as occurred can be attributed only to depletion of the water film.

Discussion

The significance of soil water for the water film inhabitants, and for the animals with a permeable integument, is not in doubt. While these animals have variable physiological responses to water stress and widely different behaviour mechanisms to deal with water shortage, it is clear that at some stage they all require free water. The quantitative measurements needed are those of water rather than humidity, while qualitative measurements will be very necessary, particularly for small animals. It is more difficult to generalize about the arthropods with an impermeable cuticle. In the few species studied, typically waterproofed cuticles have been demonstrated (e.g. Madge 1964*a*, *b*), but Madge also cites examples of mites with permeable cuticle, and all the soil dwellers have been strongly selective in their choice of high humidity. This may, however, only be necessary for their respiration, since Madge (1964*c*) demonstrated a capacity to survive considerable water shortage at low temperatures. Block (1966) extracted one half the Mesostigmata, but only 20 per cent of the Oribatei and 18 per cent of Collembola before there was any loss in weight of the sample due to desiccation. It is not clear how much the innate predatory activity of the Mesostigmata contributed to this result, and it is probably necessary to have species by species interpretation in view of the variability of the fauna. At present there is no direct evidence concerning the effect of water film depth on microarthropods; reduction of film depth in my windtunnel experiment is confounded with interaction between the main variables, so although part of the fauna becomes mobile before serious humidity stress begins it is unwise to speculate on the importance of film depth without further evidence. The difficulties of obtaining this evidence and of measuring film depth on variable surfaces in an enclosed cavity are self-evident.

REFERENCES

ANDREWARTHA H.G. & BIRCH L.C. (1954) *The Distribution and Abundance of Animals.* University of Chicago Press.

BEAMENT J.W.L. (1945) The cuticular lipoids of insects. *J. Exp. Biol.* **21**, 115–31.

BEAMENT J.W.L. (1961) The water relations of insect cuticle. *Biol. Rev.* **36**, 281–320.

BLOCK W. (1966) Some characteristics of the Macfadyen high gradient extractor for soil microarthropods. *Oikos* **17**, 1–9.

DAVIES W.M. (1928) The effect of variation in relative humidity on certain species of Collembola. *J. Exp. Biol.* **6**, 79–96.

EDNEY E.B. (1957) *The Water Relations of Terrestrial Arthropods.* Cambridge University Press.

HOLDGATE M.W. (1955) The wetting of insect cuticles by water. *J. Exp. Biol.* **32**, 591–617.

KEMPSON D., LLOYD M. & GHELARDI R. (1963) A new extractor for woodland litter. *Pedobiologia* **3**, 1–21.

KNÜLLE W. (1965) Die Sorption und Transpiration des Wasserdampfes bei der Mehlmilbe (*Acarus siro* L.). *Z. vergl. Physiol.* **49**, 586–604.

LAVERACK M.S. (1963) *The Physiology of Earthworms.* Pergamon, Oxford.

LEES A.D. (1946) The water balance in *Ixodes ricinus* and certain other species of ticks. *Parasitology* **37**, 1–20.

LEES A.D. (1947) Transpiration and the structure of the epicuticle in ticks. *J. Exp. Biol.* **23**, 379–410.

LEIGHLY J. (1937) A note on evaporation. *Ecology* **18**, 180–98.

MACFADYEN A. (1953) Notes on methods for the extraction of small soil arthropods. *J. Anim. Ecol.* **22**, 65–77.

MACFADYEN A. (1961) An improved funnel type extractor for soil arthropods. *J. Anim. Ecol.* **30**, 171–84.

MACFADYEN A. (1962) Control of humidity in three funnel type extractors for soil arthropods. In *Progress in Soil Zoology*, ed. P.W.MURPHY. Butterworths, London. 182–8.

MACHIN J. (1964*a*) The evaporation of water from *Helix aspersa* II. Measurement of airflow and the diffusion of water vapour. *J. Exp. Biol.* **41**, 771–81.

MACHIN J. (1964*b*) The evaporation of water from *Helix aspersa* III. The application of evaporation formulae. *J. Exp. Biol.* **41**, 783–92.

MADGE D.S. (1964*a*) The water relations of *Belba geniculosa* Ouds. and other species of Oribatid mites. *Acarologia* **6**, 199–223.
MADGE D.S. (1964*b*) The humidity reactions of Oribatid mites. *Acarologia* **6**, 566–91.
MADGE D.S. (1964*c*) The longevity of fasting Oribatid mites. *Acarologia* **6**, 718–29.
MELLANBY K. (1934) The site of water loss from insects. *Proc. roy. Soc.* B. **116**, 139–49.
NEF L. (1962) The roles of temperature and desiccation in the Tullgren funnel method of extraction. In *Progress in Soil Zoology*, ed. P.W.MURPHY, Butterworths, London, 169–73.
NOBLE-NESBITT J. (1963*a*) Transpiration in *Podura aquatica* L. and the wetting properties of its cuticle. *J. Exp. Biol.* **40**, 681–700.
NOBLE-NESBITT J. (1963*b*) A site of water and ionic exchange with the medium in *Podura aquatica*. *J. Exp. Biol.* **40**, 701–11.
RAMSAY J.A. (1935*a*) Methods of measuring the evaporation of water from animals. *J. Exp. Biol.* **12**, 355–72.
RAMSAY J.A. (1935*b*) The evaporation of water from the cockroach. *J. Exp. Biol.* **12**, 373–83.
STOUT J.D. & HEAL O.W. (1967) Protozoa. In *Soil Biology* (Ed. A. BURGES & F.RAW), p. 149–95. Academic Press, London.
WIGGLESWORTH V.B. (1945) Transpiration through the cuticle of insects. *J. Exp. Biol.* **21**, 97–114.
WIGGLESWORTH V.B. (1965) *The Principles of Insect Physiology*. 6th edn. revised. Methuen, London.

MEASUREMENT OF SOIL AERATION

M.H.Martin

Botany Department, University of Bristol

SUMMARY Methods for analysing the soil atmosphere, assessing the oxygen supplying power of the soil and measuring the oxidation-reduction potential are examined. The most commonly used method is the platinum microelectrode system which produces a current proportional to the rate of diffusion of oxygen through the soil. This method is unsatisfactory in non-saturated soils. In all soils membrane-covered electrodes allow oxygen concentration to be measured and recorded over a period of days.

INTRODUCTION

Soil aeration is predominantly dependent on gaseous diffusion. Buckingham (1904) and Penman (1940) have shown that the rate of gaseous diffusion in soil is closely related to the water-free pore space or porosity. There is, therefore, an inverse relationship between soil aeration and water content, and the composition of the soil atmosphere has long been known to vary with changes in soil water content, compaction and structure. In ecological problems interest usually centres around soil aeration when oxygen supply to the organism is considered inadequate. Consequently, the majority of studies relating plant growth to soil aeration have been focused on 'wet' soils. A further contributory factor to the conditions of poor aeration of wet soils is that diffusion of gases such as oxygen through water is some 10 000 times slower than through air.

The methods of characterizing soil aeration include measurement of the composition of the soil atmosphere, assessment of the oxygen-supplying power of the soil and soil oxidation-reduction potentials.

Measurement of the composition of the soil atmosphere

Early studies of soil aeration concentrated on measuring the composition of the soil atmosphere. Methods generally involved the insertion of a sampling tube into the soil and the subsequent extraction of a gas sample which was analysed using for example a Haldane apparatus. Oxygen and carbon dioxide concentrations were usually recorded.

Reuther & Crawford (1947) extracted gas samples from stationary 'wells' installed in the soil. Russell & Appleyard (1915) used a cylindrical steel tube forced into the soil from which 25 ml samples of gas were collected after rejecting the first 20–30 ml to avoid contamination. Romell (1922) employed an open-ended sampling tube and an apparatus for micro-analysis of gas to overcome some major criticisms. Leather (1915) sealed cores of soil into a metal cylinder and collected gas samples by evacuation. The resulting sample included gas originally present in solution as well as gas from the soil atmosphere.

The majority of these methods cause considerable soil disturbance during the initial placing of the sampling device, they require comparatively large samples of gas for analysis, and contamination of the sample is difficult to avoid. The source of the extracted gas presents a further difficulty in that such large samples of gas can represent only the average composition of the soil atmosphere. The soil is a heterogeneous system and while the composition of gas contained in individual pores may vary considerably it is those immediately adjacent to the organism which are of greatest interest to the ecologist.

The question of gas analysis in soils was re-examined by Hack (1956) who developed a method of extracting gas samples from soil using a hypodermic needle with a lateral orifice. Hack showed that the composition of the gas sample collected varied with its size. A sample of size of 0·1 ml gave more variable and therefore probably more meaningful results than a sample volume of 10·0 ml.

Harley & Brierley (1953) used thin walled polythene vessels initially filled with nitrogen for collecting samples of gas from the litter layers of beech-woods. Rutter & Webster (1962) designed a probe with a porous porcelain cup to obtain samples of ground-water for gas analysis. Martin & Pigott (1965)—see also Lee & Woolhouse (1966)—used teflon-membrane windows in the base of tubes containing bicarbonate solutions to obtain measurements of the carbon dioxide status of the soil. All these methods suffer from the inconvenience that long equilibration times are necessary so that measurements of short-term fluctuations in the gaseous composition of the soil atmosphere or soil water are not possible. However Jensen, Van Gundy & Stolzy (1965) have described a modification of the membrane carbon-dioxide electrode (Severinghaus & Bradley 1958) for use in soil. This electrode consists of a combined glass and reference pH electrode bathed in a thin layer of bicarbonate solution enclosed by a gas-permeable membrane. Changes in carbon dioxide concentration exterior to the membrane are reflected by changes in the pH of the bicarbonate solution. The rapid response time of these electrodes enables continuous records to be made of short-term changes in carbon-dioxide concentrations.

Oxygen supplying power of the soil

Cannon (1922) appears to have been one of the first to point out that it is the rate of supply of oxygen and not the partial pressure in the soil atmosphere that is important to the organism. However, Hutchins (1926) was the first to develop a method to measure the oxygen supplying power of the soil. Hutchins pointed out that while partial pressure may in many cases be considered directly proportional to supplying power this is not universally true. A soil containing little oxygen per unit volume may be capable of delivering oxygen to an absorbing surface at a considerable maintained rate, while a soil containing comparatively more oxygen might become depleted near the absorbing surface because its rate of supply is low. The apparatus used by Hutchins was complex and bulky, involving the operation of 56 taps, so that no field measurements were made. Nevertheless, the concept of measuring oxygen supply power represented a step forward.

Raney (1949) calculated soil oxygen diffusion rates from changes in oxygen concentration in a chamber, in the form of a probe, which had first been flushed and filled with nitrogen at atmospheric pressure. This method was used for field measurements and indicated a general agreement between diffusion rates and average yield of vegetable crops.

Lemon & Erickson (1952, 1955) introduced the use of polarographic methods to measure oxygen diffusion rates in soil and the apparatus described by Poel (1960*a*) is essentially the same with few changes. Further description of the equipment is given by Letey & Stolzy (1964). The technique depends on the reduction of dissolved oxygen at the surface of a stationary platinum electrode made negative with respect to a suitable reference electrode. A voltage of between 0·6 and 0·9 V is usually applied between the platinum cathode and a saturated calomel reference electrode. The rate at which oxygen is reduced at the platinum electrode is measured by the current flowing through the system. With an applied voltage of 0·6–0·9 V the rate of oxygen reduction in a solution is governed primarily by the rate at which oxygen can diffuse to the electrode surface (see Fig. 1, curve A). Further theoretical considerations

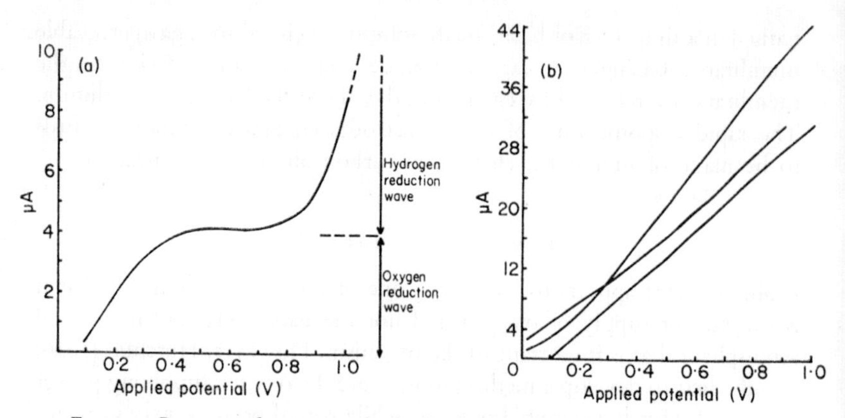

FIG. 1. Current-voltage curves obtained with the platinum microelectrode. A. Typical oxygen polarogram observed in saturated sand-water mixtures. B. Results for non-saturated clay soil. Saturated calomel reference electrode used throughout.

may be found in Kolthoff & Lingane (1952), Davies (1962), and Letey & Stolzy (1964).

Lemon & Erickson emphasize that the value of this method lies in the fact that rates of oxygen diffusion to the electrode are measured and not oxygen tensions. Thus a platinum electrode reducing oxygen at its surface is directly comparable to a plant root utilizing oxygen in soil. In addition the size of the electrode is comparable to the size of an average root.

Birkle, Letey, Stolzy & Szuszkiewicz (1964) and Van Doren & Erickson (1966) have described the effects of such factors as pH, temperature, salt concentration, electrode poisoning and soil moisture on the measurement of oxygen diffusion rates. No serious criticism of the method arises from the work of these authors so long as the electrode is used in soil of sufficient moisture content to maintain a complete moisture film around the electrode and the electrode is not left in the soil for long periods. Ingram (1964) has pointed out difficulties arising in the circuit described by Poel (1960*a*) and has suggested modifications.

Criticisms may be made that the pH of the soil could affect the reading obtained because the reduction of oxygen involves hydrogen ions. However, Davies (1962) and Van Doren & Erickson (1966) produce evidence which suggests that although the length of the plateau of the oxygen polarogram may be shortened by changes in pH, the height of the plateau remains unaltered. It would therefore seem necessary to determine the voltage giving the best current plateau for the particular soil under investigation. However, it has been found that current-voltage curves show plateaux only in saturated media (Fig. 1, curve B). Sims & Folkes (1964) demonstrated that the presence of gas phase near the water-film surrounding the electrode has the effect of lowering the half-wave potential of the hydrogen reduction reaction which then merges with the oxygen wave. Thus in drier soils where a gas phase is present, as well as aqueous and solid phases, it becomes impossible to measure oxygen diffusion rates alone and a major part of the so-called oxygen diffusion measurements is often attributable to the hydrogen reduction reaction.

Despite these criticisms, numerous investigators have shown close agreement between plant growth and soil oxygen diffusion rates—see, for example, Poel (1960*b*, 1961), Stolzy & Letey (1964). As mentioned above the performance of the platinum micro-electrode polarograph is most satisfactory in wet soils, under which conditions oxygen deficiency is most likely to occur with resultant effects on plant growth.

When using this apparatus it is advisable to select a suitable applied potential with reference to current-voltage curves for the soil under investigation. The applied potential of 0·80 V recommended by Poel appears in most cases to be at the upper limit for oxygen reduction and a potential of 0·65 V is usually more satisfactory. In soils where no oxygen plateau is demonstrable, results obtained with the platinum micro-electrode should be treated with caution. As many measurements as possible should be made for each site to allow for the heterogeneous nature of the soil and the small size of the electrodes. Van Doren & Erickson suggest that 40 determinations give a reasonable accuracy.

Willey & Tanner (1961, 1963) described a membrane-covered polarographic electrode for use in soil. This has the advantage of being temperature compensated and enables a continuous record of soil oxygen concentrations to be made over a period of about 6 days. The incorporation of a membrane (0·001 in. polystyrene film used by Willey & Tanner) covering the electrode surface isolates the electrodes from the soil; thus adverse effects of electrode poisoning, soil pH, soil conductivity, soil moisture and poor soil-electrode contact which affect the bare platinum micro-electrode method are eliminated. Interference from the hydrogen reduction wave is also eliminated and hence such membrane covered electrodes can be used irrespective of the soil moisture content. In fact this type of electrode enables oxygen measurement in gaseous and liquid phases but, as in the rotating platinum electrode described by Ingram (1964), oxygen concentrations are measured rather than oxygen diffusion rates.

Oxidation-reduction potentials

Although soil oxidation reduction (or redox) potentials are often used to assess soil aeration their interpretation is sometimes difficult. Oxidation reduction potential is a measure of the tendency of electrons to be emitted from or accepted by elements capable of changing their valency state. The emf is usually measured with a platinum electrode and a suitable reference electrode, thus giving a measure of redox potential in volts. Redox potentials are dependent on the relative concentrations of oxidised and reduced ions present and give information about only the intensity of oxidation or reduction and not about the amounts of oxidised or reduced ions present.

The work of Pearsall (1938), Pearsall & Mortimer (1939) and Starkey

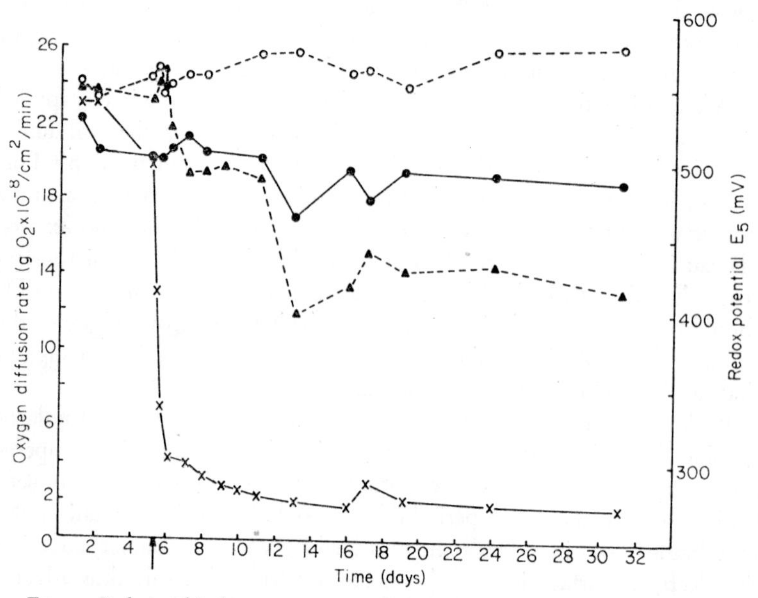

FIG. 2. Relationship between oxygen diffusion rates and redox potentials for two large pots of originally identical soil. Soil A was waterlogged with distilled water on day 5 (see arrow); soil B remained free draining throughout.

—×— oxygen diffusion rate soil A,
—●— oxygen diffusion rate soil B,
- - ▲ - - redox potential soil A,
- - ○ - - redox potential soil B.

Redox potentials are corrected to standard normal hydrogen electrode and to pH 5.

& Wight (1945) are examples in which redox potentials have been used to distinguish between oxidizing and reducing soils.

Andreason (1952) suggests that redox potentials may be used as a measure of the oxygen status of greenhouse soils. Pierce (1953) found reasonable correlations between dissolved oxygen and redox potentials of ground water. Scott & Evans (1955) however report that the only consistent relationship between redox potentials and dissolved oxygen content of soils was that low potentials developed in the absence of oxygen, and a rapid rise in potential occurred when dissolved oxygen was added to a reduced soil. In addition Scott & Evans found that only after oxygen disappeared did redox potentials decrease. Figure 2 also demonstrates this. Thus, as Patrick (1966) pointed out, redox potentials are of particular use in characterizing the aeration of waterlogged soils because in these soils oxygen is usually absent. The extent of the oxygen debt in a waterlogged soil cannot be measured directly by redox potentials. However Starkey &Wright (1945) measured the amount of reduction (oxygen debt) by potentiometric titrations of the soil with potassium permanganate solutions.

Discussion

The choice of method for measuring soil aeration for any particular investigation will largely depend on the aims and conditions of the experiment. For many purposes the membrane covered electrodes which enable continuous recording of oxygen and carbon dioxide concentrations will be most useful. Meanwhile the oxygen diffusion apparatus will continue to be a convenient method provided it can be demonstrated that the oxygen wave and the hydrogen wave are distinct.

Establishing a connection between the ecology of an organism and soil aeration may not be difficult but in poorly aerated soils it does become difficult to establish which of the conditions associated with poor aeration may be limiting the organism's performance. Not only has the effect of oxygen deficiency and the usually associated high concentration of carbon dioxide to be assessed but also the fact that the process of reduction in oxygen deficient soils may lead to toxic concentrations of substances such as ferrous iron, manganese and sulphide. Tolerance of plant roots to high concentrations of carbon dioxide seems to depend on the amount of oxygen present (Chang & Loomis 1945) and because of such interrelations it becomes difficult to study specific effects in isolation, assuming that such studies are valid.

In addition, the measurement of soil aeration may have different implications depending on the type of organism under investigation. There are numerous papers, for example Brown (1947), van der Heide, de Boer-Bolt & van Raalte (1963), and Armstrong (1964), which describe evidence for the transport of oxygen in higher plants from aerial parts to the roots. Hence the comment by Conway (1940) that the roots of plants growing in wet soils may be dependent on an internal supply of oxygen is of considerable relevance. The evidence so far available suggests that the ability of plants to withstand conditions of poor soil aeration is related to the extent to which oxygen can be obtained from internal supply. Also response to poor soil aeration appears to vary according to the species; some are tolerant of wide ranges of soil aeration while others are less tolerant and seem restricted to soils with specific aeration characteristics.

REFERENCES

ANDREASON R.C. (1952) Soil oxygen evaluation by means of redox potentials. *Florists' Review* **10**, 28–9.

ARMSTRONG W. (1964) Oxygen diffusion from the roots of some British bog plants. *Nature, Lond.* **204**, 801–2.

BIRKLE D.E., LETEY J., STOLZY L.H. & SZUSZKIEWICZ T.E. (1964) Measurement of oxygen diffusion rates with the platinum microelectrode. II. Factors influencing the measurement. *Hilgardia* **35**, 555–66.

BROWN R. (1947) The gaseous exchange between the root and the shoot of seedlings of *Cucurbita pepo. Ann. Bot., N.S.* **11**, 417–37.

BUCKINGHAM E. (1904) Contributions to our knowledge of the aeration of soil. *Bull. Div. Soils US Dep. Agric.* **25**.

CANNON W.A. (1922) Root growth in relation to a deficiency of oxygen or an excess of carbon dioxide in the soil. *Carnegie Inst. Washington Yearbook* 1921 **20**, 48–51.

CHANG H.T. & LOOMIS W.E. (1945) Effect of carbon dioxide on absorption of water and nutrients by roots. *Pl. Physiol.* **20**, 221–32.

CONWAY V.M. (1940) Aeration and plant growth in wet soils. *Bot. Rev.* **6**, 149–63.

DAVIES P.W. (1962) The oxygen cathode. In *Physical Techniques in Biological Research.* Vol. IV. *Special Methods.* (Ed. W.L.NASTUK) pp. 137–79. New York.

HACK H.R.B. (1956) An application of a method of gas micro-analysis to the study of soil air. *Soil Sci.* **82**, 217–31.

HARLEY J.L. & BRIERLEY J.K. (1953) A method of estimation of oxygen and carbon dioxide concentration in the litter layer of beechwoods. *J. Ecol.* **41**, 385–7.

HEIDE H. VAN DER, BOER-BOLT B.M.DE & RAALTE M.H.VAN (1963) The effect of a low oxygen content of the medium on the roots of barley seedlings. *Acta bot. neerl.* **12**, 231–47.

HUTCHINS L.M. (1926) Oxygen-supplying power of the soil. *Pl. Physiol.* **1**, 95–150.

INGRAM H.A.P. (1964) Examination of soil oxygen by polarographic methods. *Nature, Lond.* **202**, 1312–3.

JENSEN C.R., VAN GUNDY S.D. & STOLZY L.H. (1965) Recording CO_2 in soil-root systems with a potentiometric membrane electrode. *Proc. Soil Sci. Soc. Am.* **29**, 631–3.

KOLTHOFF I.M. & LINGANE J.J. (1952) *Polarography*. Second edition. New York.

LEATHER J.W. (1915) Soil gases. *Mem. Dep. Agric. India, chem. Ser.* **4** (3), 85–134.

LEE J.A. & WOOLHOUSE H.W. (1966) A re-appraisal of the electrometric method for determination of the concentration of carbon dioxide in soil atmosphere. *New Phytol.* **65**, 325–30.

LEMON E.R. & ERICKSON A.E. (1952) The measurement of oxygen diffusion in the soil with a platinum micro-electrode. *Proc. Soil Sci. Soc. Am.* **16**, 160–3.

LEMON E.R. & ERICKSON A.E. (1955) Principle of the platinum micro-electrode as a method of characterizing soil aeration. *Soil Sci.* **79**, 383–92.

LETEY J. & STOLZY L.H. (1964) Measurement of oxygen diffusion rates with the platinum microelectrode. I. Theory and equipment. *Hilgardia* **35**, 545–54.

MARTIN M.H. & PIGOTT C.D. (1965) A simple method for measuring carbon dioxide in soils. *J. Ecol.* **53**, 153–5.

PATRICK W.H., jun. (1966) Apparatus for controlling the oxidation-reduction potential of waterlogged soils. *Nature, Lond.* **212**, 1278–9.

PEARSALL W.H. (1938) The soil complex in relation to plant communities. I. Oxidation-reduction potentials in soils. *J. Ecol.* **26**, 180–93.

PEARSALL W.H. & MORTIMER C.H. (1939) Oxidation-reduction potentials in waterlogged soils, natural waters and muds. *J. Ecol.* **27**, 483–501.

PENMAN H.L. (1940) Gas and vapour movement in the soil. II. The diffusion of carbon dioxide through porous solids. *J. agric. Sci., Camb.* **30**, 570–81.

PIERCE R.S. (1953) Oxidation-reduction potential and specific conductance of ground water: their influence on natural forest distribution. *Proc. Soil Sci. Soc. Am.* **17**, 61–7.

POEL L.W. (1960*a*) The estimation of oxygen diffusion rates in soils. *J. Ecol.* **48**, 165–73.

POEL L.W. (1960*b*) A preliminary survey of soil aeration conditions in a Scottish hill grazing. *J. Ecol.* **48**, 733–6.

POEL L.W. (1961) Soil aeration as a limiting factor in the growth of *Pteridium aquilinum (L.) Kuhn*. *J. Ecol.* **49**, 107–11.

RANEY W.A. (1949) Field measurement of oxygen diffusion through soil. *Proc. Soil Sci. Soc. Am.* **14**, 61–5.

REUTHER W. & CRAWFORD C.L. (1947) Effect of certain soil and irrigation treatments on citrus chlorosis in a calcareous soil. II. Soil atmosphere studies. *Soil Sci.* **63**, 227–40.

ROMELL L.G. (1922) Luftvaxlingen i marken som ecologisk faktor. (Die bodenventilation als okologisher faktor). *Meddr. Skogsfors. Vaes.* **19**, 125–359.

RUSSELL E.J. & APPLEYARD A. (1915) The atmosphere of the soil: its composition and the causes of variation. *J. agric. Sci., Camb.* **7**, 1–48.

RUTTER A.J. & WEBSTER J.R. (1962) Probes for sampling ground-water for gas analysis. *J. Ecol.* **50**, 615–8.

SEVERINGHAUS J.W. & BRADLEY A.F. (1958) Electrodes for blood pO_2 and pCO_2 determinations. *J. appl. Physiol.* **13**, 515–20.

SCOTT A.D. & EVANS D.D. (1955) Dissolved oxygen in saturated soil. *Proc. Soil Sci. Soc. Am.* **19**, 7–12.

SIMS A.P. & FOLKES B.F. (1964) Limitations in established methods for the measurement of oxygen diffusion rates in soils. *Tenth International Botanical Congress Abstracts* p. 471. Edinburgh.

STARKEY R.L. & WIGHT K.M. (1945) Anaerobic corrosion of iron in soil. *American Gas Association*. New York.

STOLZY L.H. & LETEY J. (1964) Measurement of oxygen diffusion rates with the platinum microelectrode. III. Correlation of plant response to soil oxygen diffusion rates. *Hilgardia* **35**, 567–76.

VAN DOREN D.M. & ERICKSON A.E. (1966) Factors affecting the platinum microelectrode method for measuring the rate of oxygen diffusion through the soil solution. *Soil Sci.* **102**, 23–8.

WILLEY C.R. & TANNER C.B. (1961) Membrane-covered electrode for oxygen measurements in soils. *Univ. Wisconsin Soils Bull.* **3**.

WILLEY C.R. & TANNER C.B. (1963) Membrane-covered electrode for measurement of oxygen concentration in soils. *Proc. Soil Sci. Soc. Am.* **27**, 511–5

TECHNIQUES USED IN THE STUDY OF THE INFLUENCE OF ENVIRONMENT ON PRIMARY PASTURE PRODUCTION IN HILL AND LOWLAND HABITATS

M.B.ALCOCK, J.V.LOVETT* (Department of Agriculture)
& D.MACHIN (School of Plant Biology)
University College of North Wales, Bangor

SUMMARY An experiment is described which was designed to show the effect of environment on the growth of a grass sward in two habitats in NorthWales. This experiment is used to discuss the methods available for establishing what factors of the environment have most influence on growth in the field. Soil exchange between the habitats, mathematical analysis of time replicated growth curves and growth analysis, coupled with a combination of principal component and selective multiple regression analysis provided the basis of the method. The investigation was carried out with the initial assumption that crop growth is a function of the external environment and of factors within the plant and that during growth either may exert a dominant influence. Attention was only paid to periods when the external environment was the dominant influence.

INTRODUCTION

The study of the influence of climate and weather (short term deviations from seasonal trends of climate) on plant growth has followed two main lines. The first has been a statistical study of the correlation between yield and weather and the second a study of the way one or a few weather variables influence plant growth in a controlled environment. The statistical approach has been criticized for, as an example, its inability to disentangle the infinite complexity of the environment and its application

* Now at: Department of Agronomy, University of New England, Australia.

to data on final yield when the influence of the environment on growth may be expected to vary at different stages of development (Milthorpe 1965) and the emphasis is now on the use of phytotrons. We suggest, however, that the two approaches are complementary and should be given equal attention in environmental analysis (see also McCloud, Bula & Shaw 1964). The statistical approach will indicate which are the most important weather variables. Then, with physical knowledge of the environment and information of quantitative relationships between growth and single or a restricted number of weather variables appropriate theoretical models can be developed. These can then be tested in a controlled environment programmed for the level and degree of variation of the specified variables which exist in the habitats being investigated.

In this paper we describe an experiment which we conducted to establish the identity of the weather factors which had greatest influence on the yield of monoculture grass swards established in a lowland and an upland habitat. We then use this experiment to illustrate the principles involved in such investigations. The design of this experiment, the statistical treatment and our discussion are based on the assumption that crop growth is a function of the external environment and of factors within the plant and that during growth either may exert a dominant influence on growth.

METHOD

The varieties S24 Perennial Ryegrass, S37 Cocksfoot and S51 Timothy were sown on 24 April 1964 in 3 experiments situated in a lowland and in an upland (Ffridd) habitat. Details of the habitats are given in Table 1.

TABLE 1. Details of the 2 habitats.

	Lowland habitat	Upland habitat or ffridd
Altitude	6 m	290 m
Distance between sites:	1·6 km	
Soil type:	Stony heavy alluvial soil overlying red clay, gley	Light textured stony soil, free draining overlying rock, brown earth
Depth of soil:	30 cm	25·5 cm
Soil nutrients:	Mainly influenced by fertilizer and cropping	Naturally low in pH and phosphate

In Experiment I, 4 of a total of 12 replicated plots of each variety were cut at 10 day intervals from 13 April 1965 to 25 October. Each plot, which measured 1·8 m× 5·5 m, was cut with an autoscythe every 30 days and the central area of 5 sq m was used for yield estimation. At 10 and 20 days after cutting the dry weight of the crop above cutting height was estimated electronically (Alcock & Lovett 1967). Spring growth of undisturbed swards was measured from February to May in 1965 and February to June in 1966.

Four replicates of each variety were used in Experiment II which were cut every 30 days in sequence with one time replicate in Experiment I. One 0·6 m^2 sample of herbage was cut at ground level from each plot at 15 and 30 days after cutting. These samples, which were never taken twice from the same site during the season, provided estimates of dry weight and leaf area index (L.A.I., Watson 1947). L.A.I. was estimated from the area/weight relationship of a sub-sample of individually measured leaves and the weight of leaves in the main sample. Net assimilation rate (E, Gregory 1926) was calculated for the period 15–30 days after cutting during which period a closed crop canopy was attained. It was also considered that since measurements of dry weight did not include the roots estimations of *E* in an earlier period of regrowth would be subject to considerable error and therefore could not be considered in the analysis.

In Experiment III we modified the habitats by exchanging soil between them. Four plots measuring 6·4× 4·6 m were excavated in each site to a depth of 23 cm and their soil was exchanged. A further 4 plots in each site had their soil rotovated to a depth of 23 cm to simulate the degree of disturbance occurring in the exchanged soil. Each plot was sown with S24 perennial ryegrass and subsequent measurements of herbage production were made on the central 5·6× 3·6 m.

In order that soil pH, phosphate and potash content should not limit growth or create nutritional differences between the habitats all plots were adjusted with fertilizer to an optimum level for the species grown. This was carried out in February and further soil analysis at the end of the season indicated that pH and nutrient level was maintained at a high level. Nitrogen was applied as Nitrochalk (21 per cent N) to all plots every 10 days at a rate equivalent to 22·4 kg/hectare.

Meteorological measurements were made with readily available instruments obtained from the Meteorological Office. These were placed in the soil or as near to average plant height as possible in an area central to the experimental site where the grass was kept short. Details are given in Table 2.

TABLE 2. Weather data collected in both habitats

Symbol	Measurement	Method of measurement
A	Radiation cal/cm²/day (visible wavelength)	Derived from sunshine (Penman 1962)
B	Maximum air temperature °C	Bimetallic Thermograph: continuous recording at 12·5 cm above soil surface in a Stevenson Screen. Weekly mean values used obtained by planimetry of record charts
C	Minimum air temperature °C	
D	Rainfall cm	Total per week
E	Sunshine hr	Total per week. A single Campbell-Stokes recorder on the Lowland site
F	Vapour pressure mb	Weekly mean calculated from relative humidity and air temperature
G	Relative humidity per cent	Hair hygrograph: continuous recording at 12·5 cm above soil surface in a Stevenson Screen. Weekly mean calculated as in B and C
H	Wind velocity km/hr	Cup counter anemometer. 2 per habitat at 0·6 m above soil surface. Weekly mean
I	Grass minimum temperature °C	Minimum temperature recorded during each week
J	10 cm soil minimum temperature °C	Maximum and minimum temperature recorded during each week
K	10 cm soil maximum temperature °C	
L	Soil water ml	Volume of water per volume of soil. Weekly measurement

ANALYSIS OF THE RESULTS

Figure 1 shows the average growth rate of swards cut every 30 days from April to October in Experiment I and indicates the seasonal pattern of production in the two habitats. Growth rates in March are of the undisturbed swards prior to the first cut. For all three species there are three distinct periods of growth:

Period 1: Early spring growth from March to Mid-May. During this period growth rate is increasing with time.

Period 2: A period of decline in growth: Mid-May to Mid-June. This is a commonly observed phenomena where reproductive tillers are defoliated at an early stage of development and while environmental

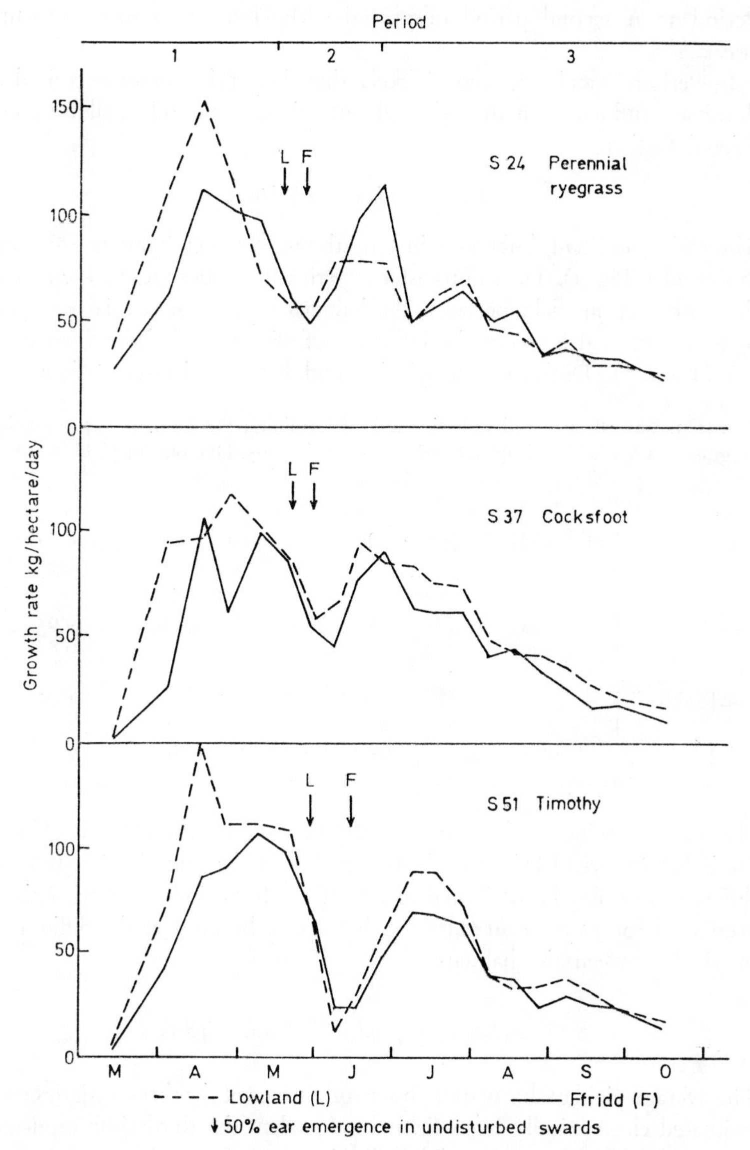

FIG. 1. The growth rate of 3 grasses in a lowland and a ffridd habitat.

factors may modify growth at this time internal re-organization of the plant is mainly responsible for the decline. This period in fact normally represents a time of high growth potential in other crops.

Period 3: A second period of growth: Mid-June until measurements ceased.

In Periods 1 and 3 it seemed likely that the environment exercised a dominant influence on growth and only these 2 Periods will be considered further.

1 *Difference between habitats*

The main consistent differences in growth rate between habitats occurred in Period 1 (Fig. 1). The soil transfer experiment (Experiment III) enabled the difference in yield between the habitats for S24 perennial ryegrass to be partitioned between the influence of climate and the influence of soil (Table 3). Differences in yield found between habitats when the

TABLE 3. The effect of soil and climate in determining the yield of S24 perennial ryegrass in a lowland and upland (ffridd) habitat in 1965. Dry weight yield '000 kg/ Hectare

Site	Soil	5 May	3 June	2 July	3 August	2 September	6 October
Ffridd (F)	F	2·93	1·77	3·02	1·59	1·49	0·94
	L	3·46	1·75	3·41	1·66	1·61	0·82
Lowland (L)	F	4·18	1·67	2·37	1·71	1·18	1·03
	L	4·39	1·86	2·34	1·81	1·24	1·06
S.E.±		0·31	0·05	0·07	0·04	0·04	0·04

plants are growing on the same soil may be attributed to climatic effects. Differences in yield between plants growing in the same habitat but on different soils may be attributed to soil effects. In Period 1 climatic effects accounted for 75 per cent (1460 kg dry matter/hectare) of the difference in yield between the habitats.

2 *Correlations of yield with weather factors*

Period 1

The average yield of dry matter above soil surface of the three varieties was estimated electronically from February to when the third time replicate was cut in May in 1965 and by clipping 0·6 m^2 samples per plot to ground level from February to Mid-June in 1966. Estimates of yield were made on a weekly or fortnightly basis and compared with average values of weather variables for corresponding periods which were themselves estimated from graphs of each variable plotted against time. Soil temperature (maximum at 10 cm) and radiation were the main factors correlated with yield (Fig. 2). While fluctuations in visible radiation are

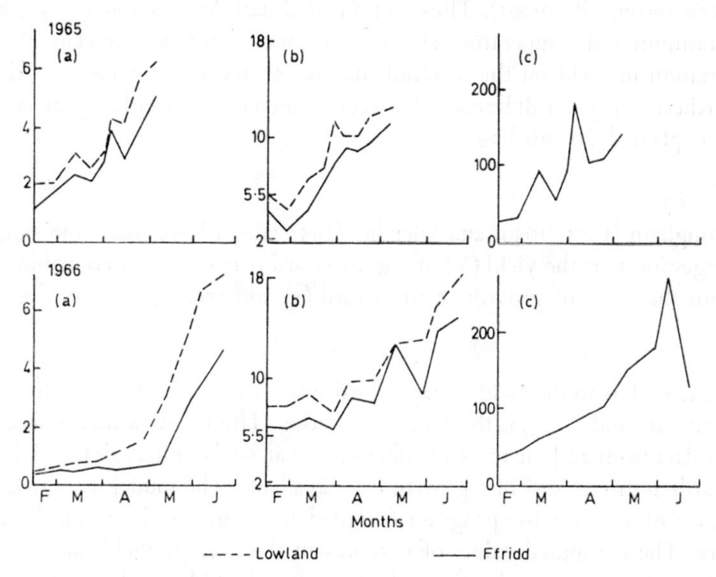

FIG. 2. A comparison of the effect of soil temperature and radiation on dry matter production in a lowland and a ffridd habitat during the spring of 1965 and 1966. (a): dry weight in thousands of Kilograms per hectare. Average of all species. (b): maximum soil temperature at 10 cm, in degrees centigrade. (c): visible radiation in cal/cm^2/day.

reflected by similar fluctuations in growth rate the habitat differences are related to differences in soil temperature since visible radiation was estimated as being similar in these closely adjoining habitats.

Growth of plants is very slow at low temperature but increases rapidly as temperature rises above a certain minimum value. Many authors have therefore suggested that variations in growth in spring may be better described by accumulating day degrees above the threshold temperature than by temperature itself (McCloud, Bula & Shaw 1964; Hogg 1965). We took the commonly used threshold temperature of 5·5°C for grassland and found that the correlation of accumulated day degrees, using maximum soil temperature at 10 cm, with yield showed a high positive coefficient. However, accumulated day degrees above 5·5°C and yield are both highly correlated with time. This correlation with time was eliminated by means of partial correlation analysis after which yield was still shown to be highly correlated with day degrees: (Lowland 1965: $r=0{\cdot}93$ $P<0{\cdot}001$, 1966: $r=0{\cdot}93$ $P<0{\cdot}001$, Ffridd 1965 $r=0{\cdot}72$ $P<0{\cdot}05$, 1966: $r=0{\cdot}94$ $P<0{\cdot}001$). Thus accumulated day degrees above 5·5°C for maximum soil temperature at 10 cm accounted for 86 per cent of the variation in yield on the lowland site and 52–88 per cent on the ffridd. Furthermore yield differences between habitats within either year could be explained on this basis.

Period 3

Brougham (1955, 1959) and Glenday (1955, 1959) have made the simple suggestion that the yield (Y) of a grass sward is made up of contributions from the stage of growth of the sward (g) and weather (w) or:

$$Y_{i(j+i)}=g_i+w_{(j+1)}+\ \ldots\ w_{(j+i)}+e_{i(j+i)}$$

where i refers to the yield at stage of growth i, j to the growth commencement date and $e_{i(j+i)}$ is the residual variation. This formula was applied to our data from 22 June to 25 October where at 10 day intervals the yield of swards at three stages of regrowth were available. The model was fitted by means of a computor program designed to minimize the residual variance. The computed values of w represent the average yield fluctuations independent of growth stage which are attributable to fluctuations in weather at 10 day intervals during Period 3. The three re-growth stages which contribute to the estimations of w were separately studied by examining the fluctuations in dry weight yield at 10, 20 and 30 days after cutting.

Values of w, and yield at 10, 20 and 30 days regrowth showed highly significant correlations with radiation, vapour pressure and maximum and minimum soil temperature. This does not preclude the lack of further significant relationships which may be masked by the influence of other variables. An attempt to identify these and to give indication of their relative importance by a process of selective multiple regression of w or yield on all 12 weather variables proved impossible. This was because an unstable matrix is formed as a result of the high intercorrelations that exist between the weather variables and because of the large number of these in relation to the 12 observations of w.

Principal component analysis, however, provides an alternative method of obtaining the relative influence on yield of all 12 weather variables. This was applied to the separate correlation matrices of weather data in the two habitats following the technique described by Seal (1966). The total standardized variance in each matrix is the number of weather variables involved. Of this variance the first four components accounted for 89·6 per cent and 84·1 per cent for the lowland and ffridd respectively. The components are independent of each other and represent groupings of variables indicating the basic dimensions of the data. The coefficients for the first four components are given in Table 4. The selective multiple regression of *w* and dry matter yield on computed values of the first four components was then calculated. Yield was related to weather conditions of the preceding 10 days. Component I alone consistently accounted for a highly significant proportion of the variation in *w* and yield (41–79 per cent $P < 0{\cdot}05$–$P < 0{\cdot}001$). This component had for both sites, high positive coefficients for radiation, vapour pressure, 10 cm soil maximum and minimum temperature. On the lowland site minimum air temperature

TABLE 4. Proportional weightings for the first four principal components of variability in weather data for Lowland and Ffridd 22 June to 25 October 1965

	Lowland				Ffridd			
Variable	I	II	III	IV	I	II	III	IV
A	0·42	0·04	0·18	0·22	0·44	0·04	0·05	0·04
B	0·17	0·33	−0·51	0·10	0·34	−0·13	0·37	−0·65
C	0·31	−0·30	−0·29	0·08	0·24	0·36	0·12	−0·19
D	0·02	−0·49	−0·25	−0·21	−0·05	0·45	−0·42	0·08
E	0·10	0·52	0·01	0·21	0·18	−0·41	0·32	0·30
F	0·38	−0·08	0·05	0·25	0·40	−0·08	−0·33	0·06
G	−0·29	−0·35	0·17	0·07	0·13	0·13	0·36	−0·71
H	0·20	−0·23	0·12	0·59	0·14	0·41	0·43	0·15
I	0·10	−0·15	0·67	−0·26	0·09	0·41	0·04	0·09
J	0·42	−0·19	0·08	−0·09	0·43	0·14	−0·24	−0·11
K	0·44	−0·05	0·19	−0·03	0·43	0·11	0·03	−0·03
L	−0·19	−0·20	0·16	0·60	−0·13	0·29	0·30	0·56
Per cent of total variation accounted for	38·75	25·40	13·33	12·15	36·32	25·81	12·24	9·73
Cumulative per cent of variation	38·75	64·15	77·48	89·63	36·32	62·13	74·37	84·10

also had a high positive coefficient and relative humidity a high negative coefficient while the ffridd gave a high positive coefficient to maximum air temperature, Table 4. There was a high correlation between the values given by component I for the two habitats (r=0·96). It is suggested that this component is related to the energy balance during Period 3.

In order to find out which of the weather variables was mainly responsible for the correlation of component I with w and yield, prediction equations were developed which assume a linear relationship and which were of the form shown below: (Kendall 1965).

$$Y=\left(\frac{y-\bar{y}}{S}\right)=b_1\left(\frac{x_1-\bar{x}_1}{S_1}\right)+\ \ldots\ +b_m\left(\frac{x_m-\bar{x}_m}{S_m}\right)$$

where S= standard deviation and $b_1, \ldots, b_m$ are standardized regression coefficients. The standardized regression coefficients for w and yield on each weather variable are shown in Table 5. These standardized regression coefficients are proportional to the correlation coefficients. The coefficients for radiation, 10 cm maximum and minimum soil temperature and vapour pressure in both habitats are considerably larger than any others and suggest that these factors are most important in determining yield. Of less importance was maximum air temperature on the ffridd and minimum air temperature in the lowland.

TABLE 5. Standardized regression coefficients for equations developed from component I (for explanation see text)

	w		30 day yield		20 day yield		10 day yield	
	Ffridd	Lowland	Ffridd	Lowland	Ffridd	Lowland	Ffridd	Lowland
A	+0·12	+0·16	+0·15	+0·16	+0·14	+0·12	+0·10	+0·09
B	+0·09	+0·06	+0·12	+0·06	+0·11	+0·05	+0·08	+0·04
C	+0·07	+0·12	+0·08	+0·11	+0·08	+0·09	+0·05	+0·07
D	−0·01	+0·01	−0·02	+0·01	−0·02	+0·01	−0·01	+0·00
E	+0·05	+0·04	+0·06	+0·04	+0·06	+0·03	+0·04	+0·02
F	+0·11	+0·14	+0·13	+0·14	+0·13	+0·11	+0·09	+0·08
G	+0·04	−0·11	+0·04	−0·11	+0·04	−0·08	+0·03	−0·06
H	+0·04	+0·08	+0·05	+0·07	+0·05	+0·06	+0·03	+0·04
I	+0·03	+0·04	+0·03	+0·04	+0·03	+0·03	+0·02	+0·02
J	+0·12	+0·16	+0·15	+0·16	+0·14	+0·12	+0·10	+0·09
K	+0·12	+0·16	+0·15	+0·16	+0·14	+0·13	+0·10	+0·09
L	−0·04	−0·07	−0·04	−0·07	−0·04	−0·06	−0·03	−0·04

Growth analysis

A modified growth analysis (which did not include the weight of roots) enabled a further examination to be made of the now established relationship between radiation, air and soil temperature and yield. The most likely weather factors to affect E are radiation and air temperature range and in addition an increase in L.A.I. will normally result in a decrease in E. The selective multiple regression of E on L.A.I., maximum minus minimum air temperature, and radiation was carried out. The amount of variation of E in combined habitat data accounted for by variation of the individual variables ranged from 53–64 per cent and a significant negative relationship between E and L.A.I. ($P<0{\cdot}01$) and a positive relationship between E and radiation ($P<0{\cdot}001$) was established. Temperature range accounted for little variation in E and was statistically non-significant.

The 10 cm maximum soil temperature was found to be the most highly correlated variable with L.A.I. accounting for 71–86 per cent of the variation in combined habitat data ($P<0{\cdot}001$).

DISCUSSION

The analysis of the environmental basis for difference in herbage production between two or more contrasting habitats requires two basic approaches. The first involves the application of experimental and statistical techniques which provide the basic physical data for establishing the identity of the main environmental factors correlated with yield. The second approach provides a testing procedure of hypotheses developed from field information which at the present time can only be accomplished in a phytotron. In the future and as information increases, it may also be possible to consider computor checking using simulated crop models.

In our experiment we have considered the first approach. The sampling procedure in Experiment I which involved the measurement of swards cut every 30 days and replicated in time to give a 10 day interval between three series satisfied the basic requirements for growth measurements. Firstly a relatively continuous record of production was obtained which ensured that differences in growth potential between the habitats was not missed (Fig. 1). Secondly, it made it possible to compare growth 10, 20, and 30 days after cutting throughout most of the year. This is essential if differences in growth are to be correlated with weather for, as Milthorpe (1965) and others have shown, the response of a crop to the environment may not be the same at all stages of its development. The use of more than one variety or species in such comparisons is of obvious importance. It should also be recognized that commercial varieties of grasses are a collection of different genotypes and it is necessary to use recently sown swards before divergent natural selection has occurred (Munro 1966; Charles 1966). It is clear that the growth of all plants is affected by other factors than the weather. If growth is to be correlated with weather then the effect of other factors must be made as small as possible or their effect separately assessed. We maintained the nutrient status of our soils at as near optimum as possible to reduce the effect of nutrition on growth. The analysis of the effect of soil and weather on differences in growth between the ffridd and lowland habitats suggest that while the major difference in growth was due to weather, some soil effects were still present.

The analysis of growth curves obtained in Experiment I enabled estimations of the fluctuations in growth (w) to be made every 10 days independently of growth stage. The yield at different growth stages which contributed to w was examined at 10, 20 and 30 days from re-growth. These growth parameters were related statistically to 12 weather observations by a process of principal component analysis followed by selective multiple regression to isolate the highly correlated component. With a large number of weather observations which constitute a number of highly intercorrelated factors it seems that principal component analysis provides a most useful procedure providing that the individual components are subject to logical interpretation. In our experiment, component I appeared to be related to the energy balance of the habitat and was highly correlated with growth.

This analysis, together with the examination of spring growth by a different procedure, emphasized the importance of soil temperature and radiation as being the main weather factors correlated with growth. The water parameters were little correlated with growth but this was to be expected since the 1965 season was wet and calculated soil moisture deficits were seldom large in either habitat.

The modified growth analysis provided further evidence of the mechanism of the growth-weather relationship. Differences between habitats were dependent on differences in soil temperature and the influence of this on L.A.I. This result agrees with that obtained by Hunter Grant & Legge (1964) for the effect of altitude on the growth of grass. Seasonal differences were determined both by the effect of radiation on E and soil temperature on L.A.I.

The 10 cm soil temperature is a complex factor to interpret since, particularly in the spring, it is highly correlated positively with radiation, grass minimum temperature, maximum air temperature and negatively with wind speed, relative humidity and soil water content. It therefore requires further investigation to establish the real significance of this factor and the mechanism of its effect on growth. Preliminary work with the influence of artificial shelter suggests that wind speed, particularly in spring, may be an important factor determining habitat differences in soil temperature and plant growth.

REFERENCES

ALCOCK M.B. & LOVETT J.V. (1967) The electronic measurement of the yield of growing pasture. I. A statistical assessment. *J. Agric. Sci.* **68**, 27–38.

BROUGHAM R.W. (1955) A study in rate of pasture growth. *Aust. J. Agric. Res.* **6**, 804–12.

BROUGHAM R.W. (1959) The effect of season and weather on the growth rate of a ryegrass and clover pasture. *N.Z.J. Agric. Res.* **2**, 283–96.

CHARLES A.H. (1966) Selection in grass populations. *Welsh Plant Breeding Station Rpt.* 1965, 60.

GLENDAY A.C. (1955) The mathematical separation of plant and weather effects in field growth studies. *Aust. J. Agric. Res.* **6**, 813–32.

GLENDAY A.C. (1959) Mathematical analysis of growth curves replicated in time. *N.Z.J. Agric. Res.* **2**, 297–305.

GREGORY F.G. (1926) The effect of climatic conditions on the growth of Barley. *Ann. Bot.* **40**, 1–26.

HOGG W.H. (1965) Climatic factors and choice of site, with special reference to Horticulture. In *Biol. Significance of Climatic Changes in Britain. Symp. Inst. Biol.* **14**, 141–55.

HUNTER R.F., GRANT S.A. & LEGGE G.F. (1964) The effect of altitude on plant growth. *Hill Farming Research Organization, Third Report* 57–60.

KENDALL M.G. (1965) *A Course in Multivariate Analysis.* (3rd impression) Charles Griffin & Co. Ltd., London.

MCCLOUD D.E., BULA R.J. & SHAW R.H. (1964) Field plant physiology. *Advan. Agron.* **16**, 1–58.

MILTHORPE F.L. (1965) Crop response in relation to the forecasting of yields. In *Biol. Significance of Climatic Changes in Britain, Symp. Inst. Biol.* **14**, 119–28.

MUNRO J.M.M. (1966) Hill land research. *Welsh Plant Breeding Station Rpt.* 1965, 72–73.

PENMAN H.L. (1962) Woburn irrigation, 1951–59, I, II, III. *J. Agric. Sci.* **58**, 343–8, 349–64, 365–79.

SEAL H. (1966) *Multivariate Statistical Analysis for Biologists.* Methuen, Lond.

WATSON D.J. (1947) Comparative physiological studies on the growth of field crops I. Variation in net assimilation rate and leaf area between species and varieties and within and between years. *Ann. Bot. N.S.* **11**, 41–76.

THE RELATION OF THE SENSOR TO THE RECORDING AND DATA PROCESSING EQUIPMENT

J.S.G.McCulloch

Hydrological Research Unit, Natural Environment Research Council,
Howbery Park, Wallingford, Berkshire

SUMMARY Advances in hydrology demand advances in instrumentation for measurement of water level, rainfall and evaporation (through the meteorological variables which control it, namely air temperature and humidity, solar radiation and wind). The use of low priced, low precision automatic stations replicated where necessary to minimize loss of records due to instrument failure or vandalism is described. The temporal and spatial variability of the factors being measured may be sampled more extensively by a number of low cost devices so as to provide a better overall estimate than would be obtained from a single complex and sophisticated device providing high accuracy at a single point.

The field data are recorded on magnetic tape in the simplest possible format and a more sophisticated counting and computing equipment is located at base for ease of servicing. The importance of reliability and speedy translation services for data tapes is stressed.

In hydrology, the yield of the hydrological cycle is the volume of flow from a catchment; given a sufficient period of flow records, statistical techniques might eliminate the necessity for any additional data! However the occurrence of floods and of droughts stimulates the requirement for some more quantitative physical understanding of the behaviour of a catchment and leads to studies of input to the cycle (as rain or snow), of the evaporative loss from the catchment (as evaporation from water or soil surfaces or as transpiration from vegetation) and of the various water storages in the catchment (as ground water, soil moisture, surface water stored in a reservoir, or as snow cover). Were quantitative knowledge of all these terms in the hydrological cycle available at all times to

TABLE

Variable	Sensor and associated circuits	Typical range required
Atmospheric pressure	Aneroid capsule with displacement transducer	950–1050 mb
Air temperature	Resistance thermometers in a variety of bridge circuits Thermistors, as above Thermocouple Transistor with power supply and amplifier	$-20°C-+40°C$
Wet bulb depression	Two resistance thermometers in a bridge circuit Two Thermistors, as above Thermocouple	0–12°C difference
Soil moisture	Electrical resistance block (gypsum or nylon block)	
Rate of rainfall	Photocell and counter	
Rainfall total	Tipping bucket with a variety of counters	300 mm/day
Solar radiation	Thermopile Solar cells	2 cal/cm²/min
Net radiation	Thermopile	$-1/4$ to $+1\frac{1}{2}$ cal/cm²/min
Wind direction	Resistance network giving step voltage outputs	360°
Wind run	Conventional cup coupled to an integrator	130 km in 1 hr at 2 m
Water level	Float system	Varies considerably. For river stage typically 0–5 m. For groundwater typically 0–50 m
Dissolved oxygen	Membrane covered galvanic cell	0–200 per cent of the air saturated value over temperature range 0°C–30°C
Suspended solids	Light source and 2 matched cadmium sulphide (photoconductive) cells in a bridge circuit	0–50 mg/l. and 0–10 000 mg/l.
Electrical conductivity	Cell using platinum or carbon electrodes	From low ionic concentrations up to a range where some saline infiltration has occurred in tidal reaches

the hydrologist or water resources engineer, the statistics of extremes might well be replaced by the laws of cause and effect!

The quantities which are considered desirable for measurement on a network scale are those of stage (or river level), rainfall, evaporation (or the meteorological variables which control it) and of water quality; there is a total of 14 relevant factors for which sensors capable of remote and unattended operation are known to exist (see Table). Some of these are most easily recorded as an event, such as the tip of a tipping bucket rain gauge, or the rotation of the cups of an anemometer; alternatively these and other like sensors produce signals which can be noted directly in digital form, while radiation and temperature are best recorded in analogue form. Of course development of new sensors or transducers is always practicable and might prove to be the most elegant solution to the problem of a mixed assembly of events, digital records and analogue records. A more obvious approach is to transform digital signals into analogue form so that the inputs are all analogue; this presupposes that the analogue to digital convertor in the recorder is more precise than the data being converted. As a result of a series of committees and working groups held under the auspices of the Natural Environment Research Council it was decided for hydrological purposes to standardize on analogue voltage inputs the minimum range of which would be 0–10 mV full scale: for most variables discrimination to 1 per cent (or about 7 bits) was adequate as the existing sensors were not as accurate as 1 per cent. Two quantities, stage and rainfall, lie outside this specification due to the very extended range necessary while recording to an absolute accuracy of 2 mm and 1 mm respectively. The density of rainfall stations is likely to be so much greater than that of stations for the bulk of the variables that it is probably best to accept a simple event recorder as a single channel instrument and similar reasoning can be applied to stage. However if a 1 per cent recording system is adopted, it is not difficult to use 2 or even 3 channels for a given variable to obtain the range required. This approach is likely to result in a more efficient, less complicated and hence less costly system than would obtain if the original specification were, say, for a 0·01 per cent system!

The recording medium which is most relevant for future developments is undoubtedly magnetic tape. Low power requirements and close packing of information are useful benefits and it is likely that long-term storage of data will be on ½ in. 9 track magnetic tape. There is not at the moment a suitable instrument for remote field recording in this format but 1, 2 or 4 track ¼ in. tape recorders may be used in a number of different ways. If telemetry of information is contemplated it is best to choose a one track recording system but there is no need to standardize on the format on the tape since such tapes can, after recording, be read directly into a computer through a suitable interface. Recorders currently on the market use two techniques; one records in the form of a succession of bits while the other samples a frequency modulated signal.

The Hydrological Research Unit Automatic Weather Station comprises a system of sensors which provides analogue voltage inputs to a battery operated data logger which, the manufacturers claim, satisfies the outline specification described above. The level of input is generally 0–5 V but, by means of an internal d.c. amplifier, inputs of 0–10 mV may be accepted. For sensors which give a change in resistance, a stabilized d.c. voltage level is available from the logger. The recording medium is $\frac{1}{4}$ in. magnetic tape in a convenient casette and the tape deck is a cheap entertainment type with an improved motor. During recording, the analogue voltage signals are converted to frequencies which are applied for a period of 1/10 sec to the recording head and thence to the moving tape; the record thus appears as a series of pulses in sequence along one channel of the tape. This system of analogue to digital conversion is perhaps the simplest and certainly the cheapest at present available and is relatively unaffected by changes of environmental conditions or of performance of the tape deck provided only that it moves the tape past the head! The more sophisticated counting equipment to convert the series of pulses into a format acceptable to a computer can be conveniently located at base where mains supplies and servicing are to hand. However, portable calibration and translation equipment are commercially available. The main translator provides an 8 hole punched paper tape output for the unit's small computer.

In Plate 1 the final pre-production prototype of the complete hydrometeorological station is illustrated. The sensors for wind run and wind direction flank the thermopile type solarimeter; the radiation-shielded, naturally-ventilated temperature screen houses 2 dry nickel resistance thermometers and 1 'wet bulb' nickel resistance thermometer; opposite is a simple thermopile-type net radiometer; the fibreglass tipping bucket rain gauge of the latest British Meteorological Office design, 150 cm^2 in area, is mounted at the standard height of 30 cm (1 ft) some distance from the mast. In each case the complexity of the sensor is minimized and all the additional circuitry which is necessary to convert the output to that acceptable by the logger is held in a rack inside the logger box, which itself is housed under a manhole cover in an underground pit.

PLATE 1. Hydrological Research Unit Automatic Weather Station Mark 3.

[*facing p.* 208

This placing was originally chosen to provide, for the logging equipment, the maximum security against vandals but the environment in the insulated pit is also much more equable and indeed is almost isothermal and thus overcomes any problems of temperature instability which affect the logging device. A bag of silica gel is exchanged every two weeks when the casette and batteries are replaced. The tape then contains records of 7 hydrometeorological variables plus a sequential time channel and a constant which can be used to characterize a particular station. Real time 'on' and 'off' must be noted on the casette by the observer, as no provision has been made to record it on the tape. Should the battery-wound clock fail to trigger the logger occasionally, (as indeed occurred on some early models due to a build up of dust between the clock contact positions), this failure would be detected in the time channel and average data would be substituted in the computer for the missing observations.

The system described above has been 'production-engineered' prior to marketing. Initially only standard stations of a particular type will be produced to avoid complexities of installation and of handling the data. Complete calibration of each station will be made by the manufacturer and the station will then be installed at the customer's site, the portable calibrator being used to check the correct operation of all the sensors. Should any particular sensor prove faulty or be damaged for any reason, the item can be replaced simply, although for certain channels it may be necessary to exchange also the appropriate circuit board inside the logger case. Thus the overall calibration of the station would not be altered. Routine schedules of simple cleaning, replacing drying agents and wet bulb wicks, replenishing the distilled water supply for the wet bulb and exchanging tapes and batteries have been produced and the stations should be capable of satisfactory operation without skilled attention. The first production models are being tested in pairs at a variety of sites in the United Kingdom and also in a tropical environment in Kenya. After 12 months' use it is considered desirable that the station should be replaced by a newly calibrated one and the various sensors and the logger returned to the manufacturers or their agents for a thorough servicing and recalibration. Certain components of the sensors such as rotary potentiometers may simply be written off and replaced and it may prove economic to replace other sensors such as the solarimeter, the black and white receiving surfaces of which deteriorate with time. This annual rebuild and recalibration of the station is likely to cost about 20 per cent of the capital cost and will provide an opportunity to up-date the system as improvements are made.

Initially a translation and computation service will be available at the Hydrological Research Unit at an economic cost but it is expected that the manufacturers (of the weather station, not of the logging equipment), will offer a competitive service in due course. Quality control will depend initially on the competence and experience of the programming staff but a number of facilities, such as an on-line graph plotter, has been provided to assist in detection of faults. Indeed it is likely that the first hint of defective or damaged sensors will come from the computer. Thus a speedy translation service is essential if breaks in operational life of the stations are to be minimized.

Further development of a water quality station recording stage to 1 part in 1000, plus dissolved oxygen and turbidity is being undertaken, in collaboration with the Water Pollution Research Laboratory. Line telemetry of the information is also being developed, without interference with the recording process. Due to the cost either of specially laid cables or of rental of General Post Office lines, it is not envisaged that a high proportion of stations will require telemetry facilities. However for flood warning purposes or river regulation it is entirely economic to telemeter information of stage and of rainfall.

The philosophy underlying the Hydrological Research Unit specification is rather analogous to that of the radio sonde. For devices which have to be sited in remote areas, far from human supervision but not from human interference and destruction, it is best to devise a system of the minimum possible cost consistent with obtaining adequate records for the purpose. In many cases high point accuracy is misleading and irrelevant due to the temporal and spatial variability of the factors being measured. Thus rainfall and evaporation (or the component factors thereof) are sampled over a minute area within a catchment, as compared to the total integration which streamflow provides. Hence while it is logical to obtain as accurate a measurement of streamflow as is justifiable on economic grounds, for the other variables a distribution of simple stations logging to a lower order of accuracy, will provide a better overall estimate than a single complex and sophisticated device providing high accuracy at a single point. The present climatological-type stations produce records of mean air temperature (from maximum and minimum mercury in glass thermometers), mean daily humidity (from 1, 2 or 3 observations of wet and dry bulb temperature in a naturally ventilated Stevenson screen), wind run per day and hours of sunshine. Such data provide adequate estimates of potential evaporation (Penman 1948): a modest standard of accuracy for an automatic weather station recording every 6 or 15 min throughout the day will suffice to replace such time honoured equipment. Thus the cost of sensors has been reduced to a minimum and the station as a whole has been made as robust as possible. In the event of wanton destruction each sensor is replaceable without recalibration as all are interchangeable.

Alternative logging systems exist and it may be possible to feed the output of 1 system of sensors into a different logging device. However it must be stressed that the system should be designed as a whole. A voltage input model of an oceanographic recorder is shortly being marketed. For use in its present form, special clamping clutch-type devices for rapidly varying quantities would be necessary and indeed have been developed for one sensor at least; this requirement arises due to the rather slow mechanical encoder. However this particular system has the advantage of 10 bit precision and a serial binary recording format. Yet a third, but more expensive system, has been developed for certain river authorities. This comprises at the moment a conventional mains operated data logger, with digital clock, recording both analogue and digital information in serial binary form. While this equipment does not meet the proposed outline specification for hydrological purposes, it is likely that future developments of this and other systems will be suitable for the type of field recording discussed.

Much of this paper has been devoted to the problem of obtaining, for a variety of sites all over the country, basic climatological-type information in a form suitable for automatic processing of the data from start to finish. For more sophisticated micrometeorological requirements similar systems may be adopted, although the element of research and hence of uncertainty in the data handling in such studies may increase the proportion of computer programming time. But even in the case of detailed micrometeorological studies made, for example in association with animal experiments, there is a definite requirement for routine climatological data of a synoptic type.

Finally before a specification for a logging system is laid down the all-important questions are: what quantities are to be measured, to what accuracy or precision and what is to be done with the resultant data? However obvious, this is sometimes not established before work begins although considerations of finance and availability of alternative equipment may obscure the issue. The right decision will avoid either of the extremes I have experienced. The first was a statement in a valuable scientific paper in a learned journal to the effect that 'the other 95 per cent of the data will be stored until a more speedy means of analysis, has been devised'. The second is that of a specification which was so wide-ranging that some quantities were to be recorded in sequence on a strip chart, the whole equipment being battery powered. While this concept was logical and technically feasible, the equipment was so clumsy and required such frequent journeys to replace large numbers of batteries that it was scrapped!

REFERENCES

PENMAN H.L. (1948) Natural evaporation from open water, bare soil and grass. *Proc. R. Soc.* A, **193**, 120–46.

STRANGEWAYS I.C. & MCCULLOCH J.S.G. (1965) A low priced automatic hydrometeorological station. *Bull. int. Ass. Scient. Hydrol.* **10**, 57–62.

CROP ENVIRONMENT DATA ACQUISITION

M. J. BLACKWELL
Meteorological Office, Bracknell
&
M. R. BLACKBURN
Meteorological Research Unit, Cambridge

SUMMARY Crop environment data acquisition requires sensors appropriate both to the nature of the phenomena and methods of recording and data-processing. Field phenomena are highly variable in space and time, but sensors can provide preliminary integration themselves. Sequential sampling permits time-sharing of expensive data-logging components.

Mains operated potentiometric recorders or digital voltmeters can already be used to provide paper- or magnetic-tape outputs which enable integration and other data-processing to be done by computer. Analogue computing methods may also be used to monitor derived quantities on site. Battery-operated magnetic-tape recorders will eventually meet most of the field-worker's requirements, when problems of on-site read-out and subsequent translation have been solved.

INTRODUCTION

The Meteorological Research Unit at Cambridge was formed 20 years ago (Blackwell 1963) to undertake experimental investigations in the field of agricultural meteorology. Once the initial choice of scientific objectives had been made, the first stage of each experiment was to decide on the physical parameters to be measured. Since the approach to the work was to be micro-meteorological, this meant that specially designed measuring devices had to be used. In addition there were numerous difficulties to be tackled which arose from the need to work in a crop environment.

In this country, and probably in many others, there were at that time very few manufacturers interested in the provision of the necessary instruments. Stage two was therefore concerned with the design and construction of appropriate sensors having suitable accuracy, sensitivity and response time. Experience has shown the importance of simplicity and reliability in all types of field equipment, but the degree of simplicity should be related to the system as a whole and not merely to a single link in the communication of the information about the environment, such as the initial measurement.

One feature which is characteristic of many ecological studies is the very large number of parameters which may have to be measured more or less simultaneously. Although the third stage of an individual experiment tended to be a somewhat *ad hoc* assembly of miscellaneous recorders, it soon became clear that the recording, preliminary data-processing and display of environmental data in a suitable form was a major problem in its own right. Indeed, there have been times when this aspect of our work has tended to detract from the main attack on the scientific objectives chosen for study.

Lastly comes the final stage of extracting the information from the records, and analysing the results. Strictly, our terms of reference are to discuss the main data-acquisition phase, but again it is of paramount importance to view the system as a whole, and any decisions on the form of data presentation must be taken with fore-knowledge of what will be done with the records. The reader may find it useful to refer back to Fig. 1 at each stage of the subsequent discussion of a typical data acquisition system.

Field phenomena

Returning to a more detailed treatment of each of the main stages which must be considered as inter-related elements of a unified systems approach, we must consider the special nature of the phenomena which occur in the field. These are usually highly variable in space, and also in time. The variations are partly systematic—with pronounced diurnal and seasonal patterns—but are combined with relatively large random fluctuations. Whether random or systematic, however, the spatial and temporal changes can be analysed to reveal the relative importance of the wave numbers (cycles/unit space interval) or frequencies (cycles/unit time interval) which may be encountered.

Many of the features to be considered are symptoms of the essentially turbulent regime which dominates the biosphere; others result from the characteristically irregular input of solar radiation through broken cloud or haze patches, and through canopy structures which are themselves in

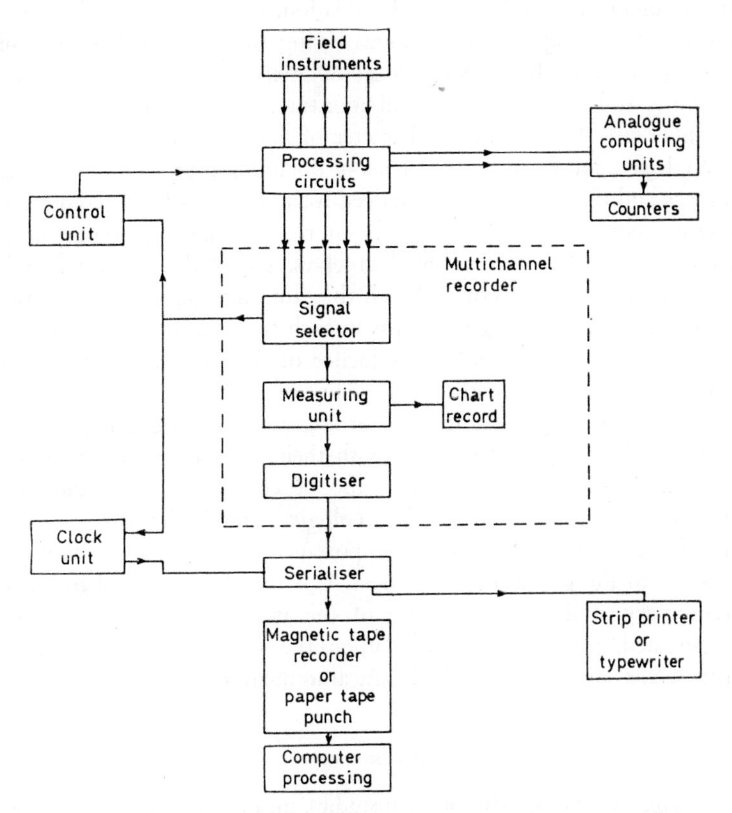

FIG. 1. General layout of equipment for field data acquisition and processing.

The diagram shows an arrangement where five transducers are being fed into a common recording system. The signals are first fed into processing circuits so that they can be matched to the mutlichannel recorder which then samples them in sequence, for measurement and recording.

The processing circuits may have more than one output, in the diagram two of the signals are being integrated by analogue computing units over relatively long periods simultaneously with the short period recording.

In addition to the conventional chart record, output may be obtained also on an electrical typewriter, or on paper- or magnetic-tape. Supplementary information (e.g. time of measurement) may be added to any of these, and a 'serializer' which accepts the information in digital form from the recorder and clock unit passes it on, digit by digit, to the output units.

Either the paper- or the magnetic-tape output may subsequently be fed to an off-line computer for more sophisticated processing.

motion. The temperature and humidity of the ambient air or of leaves, the instantaneous wind speed and insolation, the carbon dioxide concentration, are all subject to fluctuations lasting from a fraction of a second to days or years. In addition, local differences in vegetative cover or structure, soil parameters and soil moisture, provide further aspects of the inherently heterogeneous character of field phenomena.

We hear a lot about the problems of obtaining representative averages in space and time, and these are indeed formidable for both ecologist and micrometeorologist; but the latter has in recent years derived new insight from studies of the fluctuations themselves, and of the spatial variability of the parameters concerned. Much the same interest is now beginning to be engendered among ecologists in such topics as the spatial scales of distribution of species, and the influence of the time scales of radiation income on the photosynthetic efficiency of plants.

In common with the micrometeorologist, the ecologist may be concerned with mean properties, or with their fluctuations, or with both. Fluctuations of one quantity may also be correlated significantly with those of another—and from the ecological standpoint this possibility may be of major physiological importance in questions of biological response to the physical environment. It is therefore essential to make a preliminary study of the nature of the particular phenomena to be investigated, in order to decide upon the simplest design of sensing instruments compatible with the measurement requirements.

Measurement

In the context of crop environment studies, measurements can be separated into those giving a micro-meteorological picture of the physical exchanges between crop and atmosphere, and those concerned with the physiological responses at different depths within the canopy. The vigorous, turbulent mixing which occurs when the wind blows over aerodynamically rough crops creates a boundary layer whose depth is typically about one hundredth of the wind fetch. Vertical gradients within this relatively shallow layer above the crop are found to be consistent from point to point, and the meteorologist is not faced with the same need for replication as other workers in the field.

Small, delicate probes are required to follow the rapid fluctuations usually encountered in the crop environment, and may also be required for technical reasons, e.g. for temperature measurements free from solar radiation errors. More commonly, it is the mean properties which have to be measured and it is often useful to use instruments with a time-constant made deliberately large so as to eliminate the unwanted fluctuations which are then only noise confusing the desired signal. A temperature sensor having a time-constant of a minute will drastically attenuate fluctuations of period much less than a minute. This will be found an important feature when we turn to the question of sampling in time, and is really a method of partial integration.

Correspondingly, the problems of sampling in space can often be simplified by using techniques which give a measurement over an area, e.g. tube solarimeters in crops, open-path infra-red hygrometers, sonic anemometers, series-connected soil heat fluxplates. In addition to making the measurement more representative in time and space, we must consider the accuracy required. If the temperature of the ambient air is called for, we might set an accuracy of one degree as being acceptable. But as soon as wet- and dry-bulb elements are used in combination to derive humidity, we must upgrade their accuracy relative to each other to about one or two tenths of a degree. If we measure vertical profiles or gradients of temperature, we should perhaps aim at a few hundredths of a degree. In other words, even in the same situation, related measurements must be of greater accuracy than isolated readings.

Telemetry

Although, as meteorologists, we are more accustomed to thinking of sophisticated techniques for transmitting satellite, rocket or radio-sonde data, telemetry is at its simplest level concerned with instruments which measure physical quantities and convey their measurements to a distance by means of a signal (Garmonsway 1965). We have already stressed the large number of measurements required; consequently, the field worker must absorb something of the automatic techniques of the systems engineer. With this in mind, we have to make our measurements with sensing transducers which respond to the environment by varying their electrical characteristics: emf, current, resistance, or capacitance.

All such transducers can then be connected to suitable cables which bring the signals to a convenient mobile laboratory where recording, partial data-processing and display can be accommodated. Modern data-logging systems are far from cheap and it is therefore important to use the more expensive parts of the equipment to serve the maximum possible number of instruments. However, if time sharing is being contemplated, the frequency of sampling any particular channel must be matched to the time-constant of the transducer, if the mean value is not to be invalidated by fluctuations of periods comparable with the sampling interval. A hot-wire anemometer output might require to be sampled every second; a thermally damped thermometer every minute, and a soil thermometer every hour. As with accuracy and response-time it is only common sense to accept the simplest method which gives the desired result.

The last stage of conveying the transducer signals to the recording system must therefore be a sequential switching unit, capable of switching small signals down to the μV level if necessary. Implicit in this arrangement is the existence of some sort of timing unit which controls the switching cycle, and possibly other apparatus which is required to operate in synchronism. The very small signals experienced with thermocouples or photocells may require to be pre-amplified with a high stability d.c. amplifier. These have improved greatly in recent years, but it is often wise to monitor their zero stability and gain at regular intervals.

It is good practice to use as far as possible transducers which have a similar range of output, and this main group of signals is then matched by the system of recording. Signals which are too large can easily be dealt with by the use of potential dividing networks. It is perhaps worth emphasizing that, though most of the main units can be bought 'off-the-shelf', it is extremely unlikely that the much advertised package-deals will suit the detailed requirements of a particular field worker. It is even more important for many potential users of modern data-acquisition systems to realize the need for a qualified technician, to deal with the frequent faults which occur in the first year of operation, and to avoid further troubles by preventative maintenance.

Recording

Faced with the large number of parameters to be measured, and with the even larger volume of data to be processed—say 4 months continuous recording for a typical crop—it is clearly desirable to use a single system of recording. The most familiar method is to use a multi-channel potentiometric recorder: timing is controlled by mains frequency, the balancing system is compatible with the voltages which most field transducers generate, sequential switching forms an essential part of the equipment, at least 16 channels are available, sampling frequency can be changed to suit local requirements, and the chart provides an analogue display of the phenomena as they occur. This monitoring facility is often found to be invaluable in preventing the acquisition of useless data.

The span is selected to match the outputs of the majority of sensors, choosing greater rather than less sensitivity if in doubt, since attenuation is easy to arrange and provides opportunities for scaling the individual traces in direct scientific units, e.g. 100 divisions=100 mW/cm^2. On the other hand, amplification is not only expensive but liable to introduce problems of zero drift. A useful span, and the most sensitive that is available 'off-the-shelf' is 1 mV full scale. Some models provide alternative spans, choice of zero suppression and similar laboratory features, but these are not generally useful in the present context because the same facilities have to be used on all channels.

Switching commonly takes place at 5 sec intervals giving, for a 12 channel recorder, a 1 min cycle. This is a suitable sampling frequency for most purposes, and is characteristic of the general purpose ('slow') data-logger. Though designed to record d.c. voltages, the equipment can be used in conjunction with a variety of transducers. The output of non-linear devices (photocells) can be linearized by means of 'law networks'; a.c. signals (audio frequency circuits) can be rectified; and pulsed outputs (anemometers) can be passed through pulse-to-analogue rate-meters, before being recorded. Similarly, resistance bridge networks can be used in the slightly unbalanced state to provide small voltages.

Anyone who has studied a section of chart showing 16 analogue traces crossing and recrossing, with a profusion of print symbols, numbers and colours, will perhaps have wondered how such records can ever be analysed. A rough estimate of the cost of doing this manually should be sufficient to emphasize that other methods should be considered. Planimetry is almost as antiquated as visual integration of the traces, but it is now possible to trace the records with a special probe which feeds position data direct to a computer by means of a capacitative link with an underlying grid of wires. This is in fact the first mention we have made of computers but, as with the previous stages in systems analysis discussed above, the design of any one stage in the apparatus is inevitably governed by a knowledge of all the other stages.

It must be decided at the outset, therefore, whether to use a computer for some or all of the analysis, if indeed the services of a computer are readily available or can be afforded. We mean, in this context, a digital computer since the question of analogue computers will be discussed separately below. Assuming that we have already cast off our reservations about these electronic aids, we can go on to consider other ways of overcoming the problem of analysing the records. Fortunately, one of the easiest ways to eliminate the task of recovering quantified information from the multi-channel records is to fit a shaft-encoder on the balancing mechanism of the potentiometric recorder. By means of coded signals, the shaft-encoder converts each reading into digital form (000–999), and provision can be made to display these digital data in read-out form.

In order to extend the automation a stage further, the output signals can be transformed so as to operate an electric typewriter. The channels are printed out in columns on a permanent record which is easy to summate to give totals or means over suitable intervals. It is rather expensive, however, to arrange additional capacity on the typewriter to provide for automatic summation, though this can be done. If, on the other hand, computer facilities are available, it is preferable to record the binary equivalent of the digital data on paper- or magnetic-tape, and then carry out summation and any other operations by means of suitable computer programmes.

An alternative method of recording is to use a digital voltmeter in place of the potentiometric recorder. The main loss here is the analogue record for monitoring purposes, and the timing unit and sequential switch must be provided separately. The latter could, however, provide for a much higher number of channels. To shorten the scanning cycle, the switching rate could be speeded up to 1 channel/sec, giving a minute cycle for say 60 channels. A pre-amplifier would sometimes be needed because of the lower sensitivity of many voltmeters. Such a system would provide the same sampling frequency as before for any particular channel. The digital voltmeter can also be made to operate an electric typewriter or to record on paper-tape or magnetic-tape.

If it is desired to increase the number of channels still further, or to sample a number of channels simultaneously, then we move into the range of the more specialized ('fast') data-logger. The basic limitation here is often the speed of the tape reperforator which must then be replaced by a magnetic-tape recorder. In the present generation of magnetic-tape devices, it is not uncommon to have to translate the magnetic-tape record back to paper-tape, before feeding the data to the computer.

Integration

We have seen in the previous section that, where there is ready access to a computer, it is a straightforward matter to carry out an analogue-to-digital transformation of the signals and then to integrate the paper tape data by computer programme. Integration may, however, be carried out as a preliminary data-processing operation to reduce data-handling in the computer, or to render the data manageable in the absence of computing facilities.

A familiar example is the summation of anemometer pulses by electromagnetic counters. Here the rotation of the anemometer cups provides contact closures say once per revolution, and these are used to operate an electromagnetic counter via a relay. The relay provides the relatively large current required to operate the counter, but only draws a few mA through the delicate anemometer contacts. Other signals may be converted from analogue to digital form, and then integrated on counters. Most analogue-to-digital converters that are available commercially give a pulse rate output which is too fast to operate an electromagnetic counter directly. Extra equipment would therefore be necessary to step down the pulse rate by a constant factor to suit these counters, which operate up to 50 pulses/sec. However, an electronic blocking oscillator circuit, developed at our Research Unit from an earlier device by Hasler & Spurr (1958), gives a pulse rate output varying from 0–30 pulses/sec for an input voltage varying from -1 to $+5$ V.

Since the counters are basically a form of digital display, the problem is to record the information from them. Automatic photography of banks of counters at appropriate intervals has been tried, but the subsequent film processing and retrieval of data are rather time consuming. For a small number of channels we use two sets of counters and switch from one to the other so that eye readings may be taken any time during the subsequent integration period. If printing counters are used, the method can be extended to give a permanent record at regular intervals, thus providing the basis for a complete and relatively cheap system of data-logging.

Where the signal voltages have been pre-amplified to the order of a volt, it is possible to integrate them directly by means of electronic circuitry. The signals are fed into an 'operational' amplifier set up as an integrator, which provides an output voltage proportional to the time integral of the input voltage. The output voltage can be stored until required for recording purposes. This facility can be very useful in multi-channel recording systems where integration of one signal, and direct recording of another, can take place simultaneously. As soon as the stored voltage has been recorded, it may be re-set to zero and the whole operation repeated. Integrating periods with this system may vary from a few seconds to a few minutes although, with the advent of solid state circuits, longer periods will become possible.

When it is required to integrate analogue signals over periods longer than a few minutes, a magnetic amplifier and integrating motor can be used (Blackwell 1954). The amplified signal drives the integrating motor, which is geared to a pointer moving round a circular dial; the rate of movement of the pointer is linearly proportional to the input voltage.

Data processing

Data processing may be carried out at the time of acquisition of the data, as well as at a later time by computer. To reduce the amount of data obtained over a given period of time, we can carry out certain mathematical operations on the data. One example of this is the process of integration, whereby an hourly mean of a particular parameter may be obtained directly as an integrated value at the end of the period, rather than by recording several values during the period and then calculating their mean at a later stage.

A second example is the use of analogue computing elements where it is required to know the value of a derived quantity immediately after the basic data have been obtained. For instance, an on-line analogue computing system has been used by House, Rider & Tugwell (1960) to evaluate the evaporative flux of water vapour from a cropped surface, using information from wet- and dry-bulb temperature gradient and wind gradient sensors. This arrangement had the advantage of reducing the amount of data and the time needed for analysis, but a certain amount of accuracy was lost in the process and maintenance of the equipment became a problem. This often happens if the equipment has been assembled on an *ad hoc* basis which is not unlikely with work in this field where development of electronic equipment is not the main objective.

With the advent of more advanced computer equipment and facilities it would seem relevant at this stage to consider the influence of computer services on the acquisition of data in the field. Although on-line data processing equipment gives immediate results after performing relatively simple operations on the data, it may subsequently be necessary to carry out more complex calculations on the same basic data. As an example of this, it might be thought desirable to carry out a spectral analysis of the time variation of a given parameter. This would require a lengthy sequence of data, and the calculations are sufficiently complex to make the use of a computer virtually essential. Computer processing can replace some of the electronic equipment on the site, especially in the provision of supplementary data; for instance time information, which can be provided by an electric clock with digital output alongside the basic data, can equally well be provided by computer, given the time at the start of the data run and the time interval between data points.

Conclusions

The time is fast approaching when a sufficiently varied range of recording equipment will be available, commercially, to meet the needs of the field worker. Selection of any particular system should be guided by the nature of the outputs from the sensing transducers which best meet the problems of measurement. These outputs should be sampled at intervals which are compatible with the effective time-constants of the sensors, or of the phenomena observed. However, if this frequency of sampling involves the production of more data than are required in the analysis stage, or than can be accommodated on the recording tape, then integrating methods should be used as a preliminary stage of data processing before recording the mean values over a suitable period.

Where mains supply is available, there are relatively few problems remaining. The potentiometric recorder provides an analogue display for monitoring purposes, and paper-tape can be inspected visually or used to feed an on-line typewriter. Standard $\frac{1}{4}$ in. magnetic-tape is beginning to replace paper-tape as a recording medium and has a much larger capacity for data storage. For greater numbers of input channels, the digital voltmeter is more suitable.

For a fully portable, battery operated system, a choice can be made from the growing number of digital magnetic-tape recorders. Ultimately, after a period of further development, these may well become the closest approximation to the ideal specification for a field data logger. For the present, the field worker must rely heavily on expert technical assistance in order to maintain a reasonable degree of serviceability, he will not have an analogue read-out facility at the point of recording, and he will have to acquire or hire translating equipment in order to convert the field record to computer-compatible tape.

ACKNOWLEDGEMENT

This paper is published with the permission of the Director-General of the Meteorological Office. Thanks are due to Mr R.H.Collingbourne for presenting this paper at short notice.

REFERENCES

BLACKWELL M.J. (1954) Report on the automatic integration of solar radiation at Kew Observatory. *Met. Res. Ctee.* MRP No. 862.

BLACKWELL M.J. (1963) Agricultural Meteorology at Cambridge. *Mem. Camb. Univ. Sch. Agric.* **35**, 1–5.

GARMONSWAY G.N. (1965) *The Penguin English Dictionary*. Penguin Books.

HASLER E.F. & SPURR G. (1958) Light intensity meters for local and remote indication. *Electron. Engng* **30**, 690–6.

HOUSE G.J., RIDER N.E. & TUGWELL C.P. (1960) A surface energy-balance computer. *Quart. J. R. Met. Soc.* **86**, 215–31.

RADIOTELEMETRIC TECHNIQUES IN ECOLOGICAL STUDIES

J. BLIGH & S. G. ROBINSON
A.R.C. Institute of Animal Physiology, Babraham, Cambridge

SUMMARY This paper consists of a brief summary of (a) the circumstances in which radiotelemetric techniques might be of advantage in ecological studies, (b) the types of systems which can be employed in different circumstances and (c) the operational factors which must be assessed and specified before selecting commercially available equipment or designing special equipment. Comment is also made upon the problem of matching the accurate locational and physiological information about unrestrained animals, which radiotelemetry makes possible, with equally precise information about the animals' environments.

The relatively new technique of radiotelemetry offers to students of animal physiology and ecology the possibility of obtaining hitherto inaccessible information concerning the precise location, the immediate environment, and the physiological status of unrestrained animals in their natural environment. It is predictable that during the next decade, as various technical problems are solved, radiotelemetry will play an increasingly important role in ecological research in those circumstances and locations where the laws relating to the use of radio channels permits its use.

The term 'radiotelemetry' describes the transmission of information from the sensor element to the measuring system by means of a wire-less radio link. The honeymoon phase, when radiotelemetry was employed indiscriminately without regard to its suitability to the task is now over. Even the simplest radio system is an added complication in the link between sensor and recorder and most of those workers with experience in its use would counsel that radiotelemetry should be employed only when orthodox cable linkage is impossible or unsatisfactory.

There are three broad categories of linkage between sensor and recorder

The recorder unit may be in close proximity to the sensor, so that the connections are short and unexposed (Fig. 1a). The recorder may be placed some way away from the sensor, with a connecting cable (Fig. 1b), or there may be no visible link between sensor and recorder (Fig. 1c). There are, of course, other wire-less means for transmitting information, and both sound and light have been employed in special circumstances, but radio is the more versatile system, and only that is discussed here.

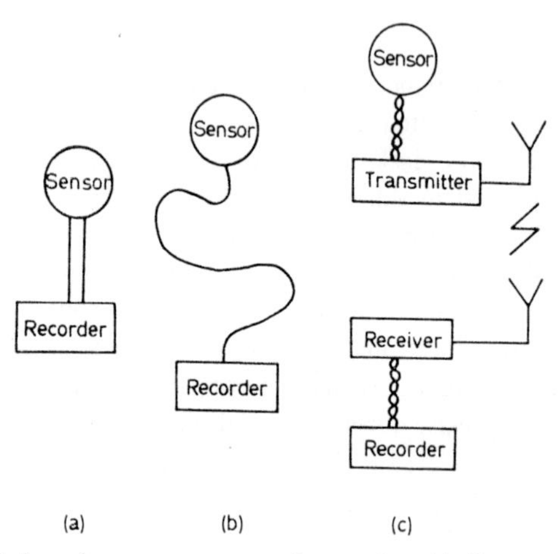

FIG. 1. Relations between sensor and recorder. (a) Sensor and recorder constructed as a unit with a fixed and close relation between them. (b) Sensor at a distance from the recorder and connected to it by a flexible cable. (c) No cable connexion between sensor and recorder. Information transmitted from sensor by radio.

The choice of the link between sensor and recorder may depend upon several considerations which will include: (i) whether the relative positions of sensor and recorder are fixed or variable, (ii) whether the locality of the sensors can be approached during the course of the observations without affecting the observations, and (iii) whether a connecting cable would be liable to damage, or would present other technical problems.

If, for example, information is required concerning the skin temperature of a free-roaming animal, or of the micro-climate close to the outer-surface of such an animal, it could be collected by one of two ways: the recording system might be carried on the animal so that the distance between the sensor and the recorder would be both small and fixed, or the sensor on the animal could be linked by radio to a fixed receiver/recorder installation.

The first alternative is subject to practical limitations some of which are inherent and others which may be soluble but persist at present. A serious inherent limitation is the unavailability of the record, even to check its existence, until the animal is recaptured. An existing, but soluble limitation is the performance of miniaturized recorders. Such instruments as are available are not suited to the physical stresses of being animal-borne, and will operate continuously for only a few hours. It should be possible to produce a discontinuously recording instrument which might run for 2 or 3 days without attention.

The use of a radio-link in such circumstances would have several advantages: (i) a small, light and rugged transmitter which will operate continuously for days, or intermittently for weeks or months, can be designed and constructed. (ii) The weight, size, ruggedness and power supply of the recorder system becomes considerably less critical. (iii) The data is immediately available if required, and the quality of the record can be determined during the course of the observation. (iv) The time during which the recorder will function without attention is of no great consequence save in terms of convenience to the operator.

While radiotelemetry will be employed most frequently in agricultural and ecological studies involving ambulatory organisms in a natural environment, or in one in which they are husbanded, its application is not restricted to such circumstances. It has also been used in the collection of climatic information from locations which are at a distance from the operator's base, and not readily accessible, and in circumstances where the attendance by an experimenter, even just to change recorder charts, would disturb the conditions pertaining to his observations.

Instrumentation

It is not possible in the available space to discuss all the considerations which may be involved in the choice or design of a radiotelemetric system. Some multi-channel systems are now offered by manufacturers. Frequently these have been designed for particular applications and have defined operational specifications. It is therefore necessary for the potential employer of radiotelemetric instrumentation to specify the performance that he would require before deciding upon a commercially available system, or embarking upon the construction of a system tailored to his own needs. The following comments can achieve little more than to draw attention to the factors which must be taken into account.

Reliability

The transition from the laboratory prototype to its practical use in the field can be a painful and time-consuming experience. The equipment must be suited to the task it is intended to perform. Additional to electronic considerations, the equipment must be designed to withstand transport and use under field conditions, and must be weatherproof. The designer has a responsibility to engineer his equipment with a view to its use by people who are concerned not with its design, but with its application. He must aim to produce an instrument into which the sensor elements can be simply plugged, and which will then require the minimum of adjustments. However, it is important that the limitations of the equipment should be well understood by the operator.

Accuracy

The accuracy required of the system must be specified at the design stage. Ideally this should equal that which would be considered acceptable with an orthodox recorder system. As this cannot always be achieved at present, the integrity of the recorded information must be determined under field conditions.

Range

The required range of effective transmission must be specified. This is a compound of the power of the transmission and the nature of the terrain, while aerial design and wavelength will also exert some influence. Range will be maximal when the terrain between the transmitter and receiver aerials is open and level, and minimal when the terrain is undulating and rocky or forested.

For some studies, particularly those related to agriculture, quite modest ranges of perhaps a few hundred yards from the receiver aerial are adequate. In other studies ranges of 20 miles and even of 200 miles have been required, and the design has accommodated these requirements.

Duration

This, also, is a matter of specification and design. It will be proportional to the capacity of the power source, and inversely proportional to the power of the transmission. If range and permissible weight of the transmitter unit have been specified, and batteries with the highest storage capacity to weight ratio have been selected, duration is essentially fixed. However, duration can be greatly increased by using discontinuous transmission. The 'on' and 'off' periods can be operated by a clock in

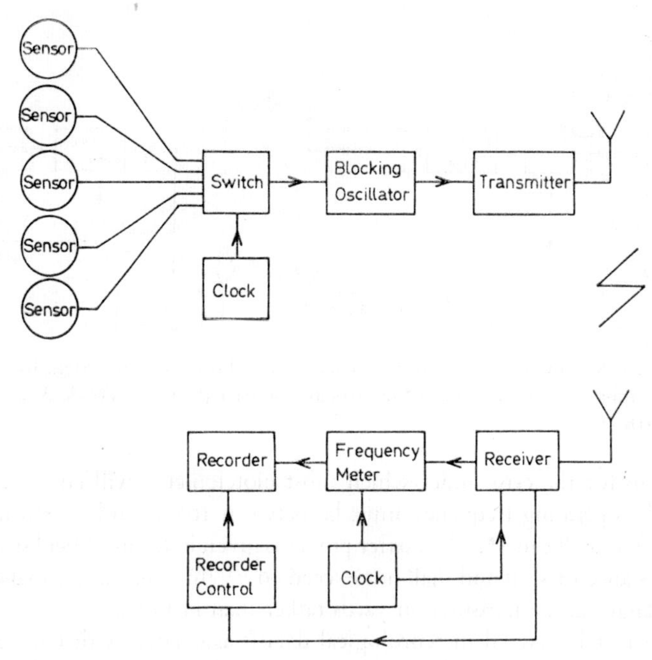

FIG. 2. Schematic diagram of a multichannel radio-telemetric system for the transmission and recording of micro-climatic data (from de Vos & Anderka 1964).

the transmitter unit (Fig. 2), or the transmitter can be switched on and off by a radio signal from the receiver station (Fig. 3).

Wavelength and interference

Selection of wave-length will depend upon legal restrictions, interference, and operational considerations.

The avoidance of interference both to other transmissions and by other transmissions is essential. Depending upon the density of band-use in the operational area, the likelihood of interference to others will increase as the power of transmission increases, and interference by others will increase as the sensitivity of the receiver increases.

Legal restrictions obviously vary from state to state. G.P.O. regulations fix both the permitted wavelengths and the permitted power output throughout Great Britain. Specifications for frequency allocation for medical and biological telemetry have now been drawn up, and potential operators should acquaint themselves with these regulations. The speci-

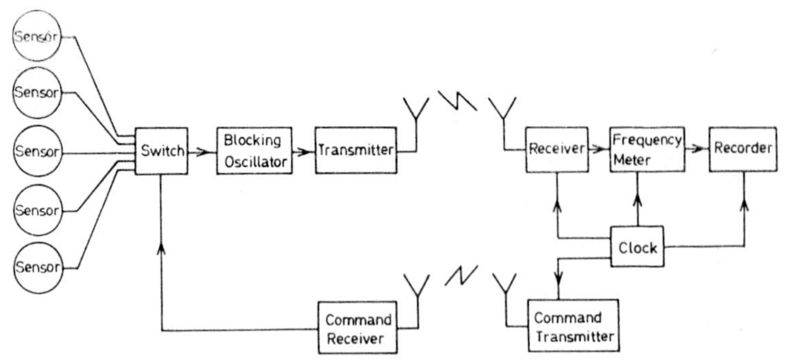

FIG. 3. Schematic diagram of base-controlled radio-telemetric system for the transmission and recording of micro-climatic data (from de Vos & Anderka 1964).

fications for the class under which most biotelemetry will come require that the operating frequency must lie between 102·25 and 102·30 megacycles/sec at all times*. The carrier power is given in terms of field strength at a distance of 50 ft and shall not exceed 10 μV/m. This means that effective range can be measured in yards rather than in miles.

Where telemetered meteorological data is associated with telemetered biological information, the system may well be operating within these restrictions. In other circumstances, where radiotelemetry is employed because of difficulty of access to the sensor elements after their installation (e.g. in a study of the micro-climates of breeding places such as birds nests), and is not to be accompanied by transmitted information about the animal, bigger transmitter units may be employable. Data links using the approved and more powerful communication channels may then be permissible. Transmission over a distance of several miles then becomes possible.

Where a choice of wavelength is permitted, the nature of the terrain between the transmitter and receiver aerials may indicate a preference. For example: the wavelength of a 102 megacycles/sec transmission is of the order of 2·9 m, which is sufficiently short for it to be deflected from obstacles such as trees. In location studies this can give rise to spurious position information. In some such studies longer wavelength transmission has been preferred as this will pass through such obstacles without serious deflection.

* The frequency has been slightly varied in: Post Office Engineering Department Draft Specification W 6803.

Parameters

The kinds of information which can be transmitted by radio is limited only by the availability of a suitable sensor unit which can convert variations in the quantity which is of interest into an electrical variation which can be used to modulate the carrier signal. The majority of the meteorological parameters is amenable to this treatment. Wind direction, wind velocity, temperatures, and radiation load are readily expressible as electrical variations. Changes in humidity, and the incidence of rainfall may require the development of suitable sensor systems but do not appear to present insuperable difficulties. Information concerning several physiological parameters have been successfully telemetered. These include body temperatures and pressures, biogenic electrical potentials and respiratory frequency. Attempts are now being made to devise sensor units which will detect and transmit information about chemical changes.

The simultaneous transmission of several pieces of information, greatly complicates the radiotelemetric instrument. The possible solutions are discussed in the next section.

Multi-channel systems

A single piece of either physiological or meteorological information is of only limited interest by itself, and more generally the system will be required to transmit several pieces of information. Their simultaneous transmission would either greatly complicate the instrumentation, or require as many transmitters, receivers and separated wavebands as the pieces of information. Such systems would only be considered if truly continuous records were imperative, and truly simultaneous comparisons need be made between the different parameters. For many studies this would be a quite unnecessary complication.

There are several means by which a number of pieces of information can be conveyed on one radio carrier. Each has merit in particular circumstances. Three systems which might be employed for the transmission of meteorological information are:

(a) *The standard F.M.–F.M. system* in which each channel controls the frequency of a sub-carrier oscillator. Here the signals are combined and transmitted continuously.

(b) *Pulse modulated systems* in which the information channels are sampled in turn at a rapid rate and the instantaneous value of each channel in turn modulates one of the characteristics of a transmitted pulse (e.g. its duration). A reference signal from the transmitter ensures that the receiver remains in synchrony. Information is thus transmitted in a quasi-continuous manner.

(c) *Slow scanning systems* in which the information channels are sampled sequentially at a slow rate. With such systems each item of information may be recorded for several seconds at intervals of 1, 2, or more min, depending upon the number of channels to be sampled and the duration of each sample.

With each of these systems data can be collected at intervals of, say, 1 hr thus conserving battery power and recorder chart.

The transmitter-receiver unit for micro-climatic data constructed by de Vos & Anderka (1964) is a slow scanning system (Fig. 2). Two synchronized clocks were used, one controlling the switching of the sensors into the transmitter, and the other to switch on the receiver and recorder. In the field operation of this system, information was lost due to a lack of synchrony between the two timing units (Dyer, Anderka & de Vos, personal communication). This fault was obviated in a modified version of the system (Fig. 3) in which the clock in the receiver system also activates the switching mechanism and transmitter by means of a reverse-direction radio command.

Where all the sensor elements connect to the one transmitter this 2-way radio is only necessary if transmission is to be discontinuous. Where frequent sampling of the parameters is required the switching unit can be operated continuously, together with the transmitter, receiver and recorder. In this event one channel into the switch must carry an identification signal.

Systems are now being developed for receiving and recording information from several locations or animals, with time-sharing of one carrier signal wave-band and a single receiver and recorder system. Here the individual transmitter units must be interrogated in turn in response to command signals from the recorder station.

Relations between climatic and biological information

The discussion has so far centred upon the use of radio telemetric techniques for the collection of climatic and micro-climatic information. There is another aspect of radiotelemetry which needs to be considered. By the use of these techniques the biologist can now learn a great deal about the location and the behavioural and functional activities of unrestrained mammals in their natural habitats, but this mass of information can be fully appreciated only if it can be related to equally precise information about the environment. This aspect of radiotelemetry has received inadequate attention.

With what should one seek to collate continuous, or near-continuous information about an animal's movements, feeding, drinking and sleeping patterns, choice of habitat, and such physiological parameters as body temperature? Are temporal and spatial profiles of the climatic variables within the area covered by the study sufficiently informative, or is it necessary to record contemporary information about the micro-climate of the animal's environment? The first choice involves a detailed micro-climatic survey of the area using orthodox techniques. This is time consuming, but the data is useable for many studies within the same area. Adequate micro-climate profiles must await development of inexpensive reliable, portable and automatically recording micro-weather stations. The second choice, that of transmitting contemporary climatic data has not yet been attempted. There are obvious snags: firstly the sensor elements must stand off from the animal, and will, therefore, be liable to damage as the animal moves through vegetation. Secondly, the attachment to the animal of both sensors and transmitter, by means of harnesses, may so modify the animal's behaviour as to invalidate the results.

These are problems which will have to be faced in the near future. At the moment we can only indicate the directions in which the advent of radiotelemetry as a research tool will lead the environmental physiologist and ecologist, and the problem that they will then encounter in matching their biological data with that of the environment.

REFERENCE

DE VOS A. & ANDERKA F.W. (1964) A transmitter-receiver unit for microclimatic data. *Ecology* **45**, 171–2.

RECORDING SOME ASPECTS OF A FOREST ENVIRONMENT

A. I. FRASER

Forestry Commission Research Station, Farnham, Surrey*

SUMMARY The recording of any environment depends on the successful integration of the sensing instrument, the recording equipment, and the data processing system.

The paper describes the method developed by the Forestry Commission for measuring wind velocity profiles above a forest, and recording the results on a battery powered automatic data-logger.

Cup-generating anemometers are used to measure wind velocity, and the frequency modulated output from these is recorded by the logger through a pulse counting circuit. The data-logger is basically a digital voltmeter which can scan 50 channels in $6\frac{1}{2}$ sec and digitize the signal from each channel on punched paper tape.

A special computer programme has been written to check all data tapes for errors and to give an edited tape for further computing and print-out.

INTRODUCTION

That part of the atmospheric environment exploited by a tree crop is both large and complex by comparison with most other plant communities, so that special problems are encountered in measuring it.

In Britain, these problems may be further complicated by remoteness from normal working facilities, such as mains electricity and a reasonably stable environment. Indeed remoteness itself is a major handicap in the study of some phenomena, because it necessitates either a prolonged stay on site until the weather conditions are just right, or a hurried visit to the site when suitable conditions are forecast.

* At the Department of Forestry and Natural Resources, University of Edinburgh 1967–9.

Exploration of the atmosphere among or above tree crowns, some 40 to a 100 or more feet above ground presents many practical problems.

Trees because of their height are particularly susceptible to wind pressures, and the Forestry Commission have been interested for some time in the influence of wind on tree growth, and in the problem of storm damage.

Measurements of mean wind-velocity profiles above a forest plantation, require that several instruments in one array are recorded simultaneously. It is also necessary to sample the air flow at several locations in order to obtain a measure of the variation. At least five sampling heights and five locations are desirable and thus a minimum of 25 instruments are required for satisfactory measurement of wind velocity.

Two important factors must be taken into account when designing a recording system; the method of data processing that will be employed, and the calibration of the recording apparatus. The recording system that we have developed demonstrates how both these requirements can be satisfied, and may at the same time prove interesting to others who wish to record in the field.

The quantity of data that can be collected in even quite a short period from such an array of instruments dictates that it should be recorded in some easily handled form. Chart records for example, present an almost impossible task to analyse, since each chart has to be handled manually or some automatic chart reader acquired to digitize the records.

The Forestry Commission has a digital computer, so that ultimately the data must be punched on paper tape, in a suitable format for direct reading into the computer. Three alternative commercially available systems were available at the time. We decided against the construction of our own system, because of limited workshop facilities.

Possible recording system

The first and simplest system is to have anemometers (or any other sensors) which close contacts at a predetermined total, and to count the number of closures on an ordinary mechanical or electrical counter. A refinement of this system is to record with automatic print-out counters, with a time switch designed to operate the print at a predetermined interval, but these are expensive (approximately £40 each) and can only count one channel at a time. The ordinary electrically operated counters could be read by observers and the data recorded on paper. However, it was felt that several observers would be needed to read the counters sufficiently quickly, and the processing of the data would cause additional problems. It would also be necessary to be on site during the conditions that we wished to record.

The second possible system is to use a small portable battery operated magnetic tape recorder, which could sample a voltage output from generator anemometers, and record this in some digital form on the tape. Such a machine was available at the time (Limpet Logger), and the possibility of having one per station, scanning each of the instruments rapidly in turn, was seriously considered. It was said to be possible to co-ordinate the loggers in such a way that they all scanned simultaneously. The output on magnetic tape required translating into punched paper tape, which at that time proved to be a difficult operation, though it is now possible. Several systems are now in operation using these machines but at that time we were not in a position to do any development.

The third system, finally chosen, is based on a single 50 channel automatic data-logger, as commercially available from several manufacturers, recording the output from generator anemometers. Because we require to have it battery operated only the Westinghouse Brake and Signal Co. could supply a suitable machine at a reasonable price.

The ultimate refinement of such a system, of course, is to have the data-logger at base, and use telemetry to transmit the data from the sensors in the field, but this is very expensive, and the advantages could not easily be justified. We therefore decided to make provision on site to house the data-logger in a convenient location, and to run wires into it from the sensors. This did not present any great difficulties, as small wooden huts are fairly cheap, and can be readily insulated.

Details of the recording system used

The data-logger is basically a digital voltmeter, which can digitize a d.c. voltage input to very high accuracy. The number of significant figures required determines the length of time taken to digitize each channel, though it is the punch which imposes the ultimate limit on speed.

The 50 channels are scanned in turn, and the output from the analogue to digital converter (A.D.C.) is temporarily stored and then punched in binary coded decimal (the code used by the Forestry Commission Sirius Computer).

The three main types of sensors in use at the present moment, are cup-generating anemometers, potentiometers and thermistors.

The anemometers used are Met. Office Mark IV generators producing an a.c. output, whose frequency *and* voltage amplitude are proportional to the wind velocity. The frequency output is used since the line does not affect the frequency transmitted, and it is capable of much more accurate resolution.

A pulse counting frequency meter was developed for us by Mr R.F. Johnson of the National Physical Laboratory, and this gives us a suitable d.c. voltage which is directly proportional to wind velocity.

The potentiometers are useful for recording any changing phenomena which have sufficient power to drive a small lever or wheel. We have attempted to use these for recording wind direction but so far have had trouble with excessive wear, and will have to change to another system. We also propose to use potentiometers to measure and record tree deflection. The logger can supply a stabilized voltage, which is wired across the ends of the potentiometer coil, and the middle wiper arm is tapped and the voltage drop recorded.

The thermistors are also simple to use, provided the correct type is chosen to suit the voltage and power dissipation. A transistor circuit is used to provide a stabilized voltage and the logger records the voltage drop across the thermistor.

With both potentiometers and thermistors, power losses in the line are important and steps have been taken to allow for these. We have built a small trimming potentiometer into each circuit, so that we can adjust for zero errors imposed by variable length of line.

Apart from the calibration of the sensors themselves which is done in the laboratory or wind tunnel, it is essential to be able to calibrate the data-logger to check that a given input is being digitized correctly. This is done by fitting a jack plug in each input circuit, so that the transducer can be isolated, and an alternative known input substituted. With the anemometers, an oscillator is used to feed in a range of frequencies, and both zero and drift errors can be corrected. With the potentiometers and thermistors, a resistance box is used in a similar way.

The data-logger has the facility whereby any channel can be selected manually, and repeatedly scanned. There is also a visual display of the output from each channel, thus if a known input is fed into each channel in turn, the output of the A.D.C. and the punch can be checked against each other and the input. Any corrections necessary can then be made. This operation is carried out about every month to six weeks and no error greater than 1 per cent in the A.D.C. has been detected. The punch output has always agreed with the A.D.C.

We wished to be able to scan the 50 channels as rapidly as possible, in order to be able to obtain a large number of samples on each channel. A scanning rate of 8 channels/sec was eventually achieved by modifying the output format, though our need to have the system battery operated and lack of funds did make it a difficult task.

The main determinant of scan rate was the punch, since the fastest battery operated model available had a maximum output of sixteen characters per second. Had we had a large store to allow the A.D.C. to proceed at its full rate and the punch to catch up when scanning ceased, we could have achieved higher rates, but much greater expense and power requirement precluded this. A maximum output from the punch of 16 characters/sec meant that only five channels could be scanned per second if we required the output to 0·1 per cent or 8 channels/sec with the output to only 1 per cent, provided no time was wasted with spaces between channels. We accepted the latter alternative which gives us 100 digit word from each scan, followed by a 2 digit code (in fact ++) to indicate the end of the scan. All this was done with the future data processing in mind, and I will describe later the computer programme that was developed to deal with this rather unusual format.

The time required for each complete scan was therefore fixed at a minimum of $6\frac{1}{2}$ sec and this gives 9 samples/min on each channel if the logger is continuously operating. This was considered satisfactory for periods when the weather conditions were suitable for sampling at this density. However, each scan uses 10 in. of tape, so that $7\frac{1}{2}$ ft of tape/min would be accumulated.

We therefore had a selector mechanism built in to allow us to reduce quite drastically the scanning interval, with provision for automatic or remote switching to continuous logging during appropriate conditions. The scanning intervals we have chosen are every 5 min, 15 min, 30 min, 1 hr and 2 hr, though it is no great problem to change these to almost any other values. We felt, that these would give us estimates of the mean values over 1 hr, 3 hr, 6 hr and 24 hr periods respectively.

The logger itself uses 6·3 amps, and if switched on all the time, the normal heavy duty 12 V batteries would only provide enough power for about 18 hr. To conserve power, the manufacturers developed an automatic switching mechanism to turn power on and off just before and after each scan. Thus when the logger is not on continuous logging everything is switched off, and the power is only on for a total of 1 min for every scan: on 5 min intervals this has saved 80 per cent of the power consumption, and now means that the logger can operate for a week on one set of batteries.

Most of the logger is constructed with printed circuit boards which plug in to racks. The design follows a very logical system, and with the aid of drawings it is possible to carry out most repairs by simply replacing any faulty boards with spare ones. In fact in 16 months we have only had to replace three boards. Diagnosis of a fault is straightforward using a process of elimination. Since the same boards are used in several parts of the logger, it is possible to exchange similar boards and observe the result. For example a few channels, which were all connected through one board, were giving outputs which were clearly erroneous. Exchanging the common board with another similar one, corrected the error on the first set of channels and produced erroneous outputs on others, thus identifying the faulty board. These printed circuit boards are all readily accessible, and we carry a spare of each type.

The Logger comes in two steel cabinets 4 ft × 2 ft × 2 ft, which we have housed in a small wooden hut on site. We transported both cabinets in the back of a long wheel base Land Rover, but they are not suitable for frequent moving from site to site.

The output

As indicated above, the output is a 100 digit word terminated by 2 plusses and 2 spaces, and as such is not suitable for direct print-out on a teleprinter which normally only has 64 characters per line. It is important, however, on removal of the tape from the Logger to do some editing, and to check the tape for faults, and to analyse the data.

A special programme was written, which would read-in these 100 digit words; count the number of words, count the number of digits per word, and check that each character was an acceptable digit; then insert spaces and the necessary codes for carriage returns—line feeds at appropriate places, so that the data on the tape could be printed out in a convenient form by a teleprinter. The output tape from this programme can then be printed out in the normal way, and it also forms a data tape for subsequent analysis. A tape with 600 scans on it takes about 10 min on the computer for complete checking.

It is essential to have all programs required for dealing with the data ready before the data logger is commissioned, because tape quickly builds up, and backlogs are difficult to deal with. We note the time at the start of each tape, and normally write it in manuscript on the tape, though the Logger has the facility for manually punching headings if required. We also record the time when the tape ended, so that knowing the scanning interval or rate, we can check the total number of scans actually completed against the number that should have been done.

These normally agree, though on one occasion we did detect a fault in the automatic switching mechanism when the actual number of scans greatly exceeded the theoretical number.

General

The system we have adopted has been designed to fit our particular requirements, but the basic idea is so versatile that almost any requirement could be fulfilled. For example, fewer or greater number of channels could be accommodated, greater accuracy of resolution could be achieved, output could be on a teleprinter or punch and teleprinter combined. Almost any sensor can be used provided that small analogue changes in voltage are produced.

The advantages of having all data on one tape are considerable, and though all data may be lost if a serious fault develops, our experience has been that these systems are so reliable that this is rarely likely to happen. It is possible at extra expense of course to build a digital clock into the system so that the time of each scan can be logged, if this is important. The cost of our unit, is approximately £45 per channel, which compares very favourably with any other system, when one takes into account the cost of data-processing by such systems as manual punching, or translating from magnetic to punched paper tape.

Some practical considerations

The main enemy of electronic equipment is condensation, and great care has to be taken with any instruments to avoid this occurring. We have found that a wooden hut lined with insulation board on the inside, reduces temperature fluctuations to an acceptable level so that condensation has not been serious. We have also kept a tray of silica-gel in each cabinet of the logger, and have the logger standing on a platform 18 in. off the ground. These precautions have prevented condensation interfering with the working of the equipment.

We have maintained a log book with the logger since we received it, in which everything done is recorded. This has proved invaluable in testing and checking the equipment, and we have referred back to it many times to check the details of some adjustment that has been made.

We have also maintained a log book at the research station to record the progress of each tape sent in from the data logger, through the various stages of processing. This includes a full record of any faults that are found in a tape by the check programme. This has also proved a useful exercise, and has meant that many of the jobs can be done by people not directly familiar with the equipment, in the event of the usual operators being absent.

ACKNOWLEDGEMENTS

The development of the data-logger was largely carried out by Mr R.Acreman of the Westinghouse Brake and Signal Company, to whom I am very grateful for his interest in our problems. Mr R.F.Johnson of the National Physical Laboratory kindly developed the pulse counting circuits and gave much useful advice for which I am also most grateful. Finally, Mr D.Stewart of the Statistics Section of the Forestry Commission developed the program for dealing with the output from the logger and his help at all stages has been much appreciated.

PROCESSING THE OUTPUT OF PAPER TAPE RECORDING EQUIPMENT

J.K. Brookhouse
Reading University

SUMMARY This paper outlines a method of writing large or complicated computer programs. This method is illustrated by showing some of the steps in writing a program, for a small computer, to analyse the output of automatic recording equipment. Emphasis is placed on the detection and removal of errors on the paper tape. For this purpose it is necessary to record information which would not be essential if the equipment were working perfectly. The facilities that would become available if a larger computer were used are also discussed.

INTRODUCTION

To help him to complete successfully a large data recording project the biologist will need the help and advice of at least three experts, the equipment manufacturer, the statistician and the computer programming specialist. Suppose that the biologist wishes to record six characters of information per min and that the manufacturer claims that his equipment only mispunches 1 character per 10,000, then the biologist may decide to take this risk and go ahead. In fact the chance of producing a perfect record for one week is about 1/400. Unless special precautions are taken the computer program will go wrong at the first error on the tape. As only a small portion of the record will be analysed it will not be possible to answer many of the questions posed. The biologist can avoid these difficulties by consulting the statistician and programmer long before the amount and type of information to be produced has been decided. The statistician can give a good guide as to the frequency of collection of information and also suggest methods of dealing with the small amount of missing data. The programmer can suggest the information that should be put on the tape to enable the program to find its place after an error and hence salvage the rest of the record. Strictly none of this information would be necessary if the equipment worked perfectly.

In this paper I will assume that this team of experts have met. They have decided to use recording equipment that is going to generate a paper tape. This tape will be processed on an available computer. I want to consider two problems; the first is how do we instruct the computer to do what we want it to do with the paper tape record and second, what are the main features that are required of a computer program for analysing data generated by automatic recording equipment.

1 *Use of programming Languages*

If your establishment has access to an electronic digital computer then it is quite likely that you have attended a short programming course. In this paper I will not assume that you have attended such a course but I strongly recommend that anyone contemplating the use of data recording equipment does so before embarking on the project. Such courses usually consist of a series of lectures illustrating the rules of a programming language and a few tutorials in which simple programs may be written.

Almost invariably these courses introduce either FORTRAN or ALGOL, the two programming languages used by the majority of scientific computer users. These languages contain sufficient English words to be readily intelligible to human beings. They contain so few ambiguities that another computer program called a compiler can read them from suitably prepared cards or paper tape and generate an equivalent set of computer instructions. The natural redundancy in such languages enable the compiler to perform a number of syntactic checks but it cannot detect or remove logical errors in the specification of the problem.

The best programming language to use is the one preferred at the computer installation to which you have access. Writing in this language will enable you to get more help and advice than you would using any other. Both languages were designed specifically for processing numbers. This ensures that virtually all the logical and arithmetical operations that you might need are easily specified and efficiently executed. Consider the following two examples of ALGOL, first, the conversion of temperature from Fahrenheit to Celsius, and second, the calculation of the mean and standard deviation of maximum daily temperature.

```
(i) TC:= (TF−32·2)* 0·555556;
(ii) at the beginning | SDAY:=SMAXT:=SSMAXT:=0;
     for each record  | read T;
                      | if first record of day then MAXT:=T
                      | else if last record of day then
                      | begin SMAXT:=SMAXT+MAXT;
                      |     SSMAXT:=SSMAXT+MAXT*MAXT;
                      |         SDAY:=SDAY+1;
                      | end else if MAXT<T then MAXT:=T;
     at the end       | print SDAY, SMAXT/SDAY,
                      |     Sqrt ((SSMAXT−SMAXT*SMAXT/
                      |                    SDAY)/(SDAY−1));
```

These examples show the relative ease with which algebraic expressions may be written and evaluated and also the simple way in which alternative computations may be chosen.

When we are considering the problem of processing paper tape from recording equipment, both languages have a common and serious deficiency. This occurs in their input routines, which are designed to convert numbers that a human being can recognize into the computer representation of those numbers. For most problems this is ideal, however, for inter-machine communication there are several objections. It requires an expensive piece of additional equipment to enable a machine to generate information that a human being can recognize. This appears to be especially ridiculous when so much information is going to be generated that no person will have the time to look at it. Such a conversion could well increase the amount of output by a factor of 3, which will result in punch errors occurring three times as frequently in real time. The computer will almost certainly convert the information back into a very similar representation to the one from which it was produced. Even if this is not the case, the computer is ideally suited to deal with the problem of code conversion. It is capable of transforming any code into any other code providing there is a 1 to 1 or a many to 1 relation between them. Finally, the use of such special equipment is very unlikely to obviate the need for a special input routine. The standard ones almost always fail to inform the program that particular non-digital characters have been read. It is extremely unlikely that you will obtain any benefit from generating a standard paper tape and it may lead to considerable disadvantages in the programming problem.

If you decide that your equipment is to produce a simple code with error detecting and correcting information then you will need a method of reading this into the computer. This is most easily achieved by getting a programmer to write an ALGOL procedure or FORTRAN subroutine for this special purpose. By writing the name of this procedure at the appropriate place in your program a fixed quantity of information will be read from the tape to specified locations in the computer store.

2 *Flow Diagrams*

In general the biologist has had very little mathematical training and the programmer has had considerable mathematical training. The different disciplines of the two can lead to a difficulty in communication. This difficulty can be eliminated if the biologist understands how the programmer sets about the writing of a computer program. With such a knowledge the biologist will have a good idea why he is asked certain questions and can give the required answers. This will almost certainly result in a program that not only gives you the information you require, but also gives you additional information that you may have thought of as desirable but impossible to obtain.

For the small or simple problem the program will usually be sketched out on a sheet of paper and be punched onto cards or paper tape. The program will then be run with some test data. The first run will almost certainly show up some syntactic errors in the program. These are corrected and the next run or so will most likely show up some logical errors, special cases that the program fails to deal with correctly. After 3 or 4 attempts the program appears to be fully operational and can be used on the real sets of data. For the larger and more complicated programs this method breaks down completely. We may consider, for this purpose, that the processing of paper tape records is a complex problem. For such problems the programmer produces what is called a flow diagram.

Figure 1 shows a flow diagram of a program designed to process the paper tape record. The problem has been broken down into a number of logical operations each of which is represented by a box in the diagram. A brief description of the operation is written within the box. An arrow from one box to another indicates the order in which the operations are to be obeyed. An operation such as 'read record' or 'print results' is represented by a rectangular box. Such a box has one arrow entering it and one leaving it. This implies that the operation 'read record' is performed as a whole and is followed by a unique operation namely 'record complete?'. It is not possible for two logical paths through the diagram to merge within an operation box. The operation 'record complete?' is a logically different type of operation and is therefore represented by a different shaped box, one with curved ends. There is only one entry point into the box but there are two exits. The operation is equivalent to making a decision as to which of two branches of the program to perform. If the record is complete, that is free from errors, then the program processes the data otherwise it prints out some monitoring information and goes on to read the next

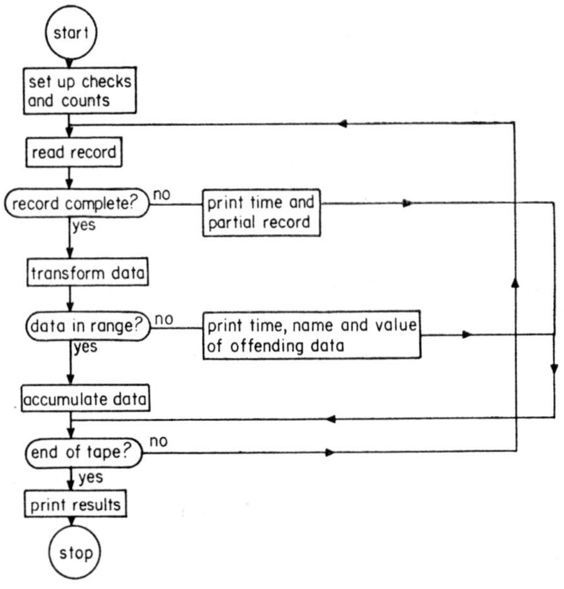

FIG. 1. Processing a single tape from automatic recording equipment.

record. This operation illustrates some of the most powerful attributes of an electronic computer. First, the ability to decide what operation to perform next in the light of previous operations. Second, the ability to go back and repeat the same sequence of operations on the next set of data. The round operation boxes are essentially labels, they indicate where the program starts and finishes. In complicated flow diagrams such labels can be used to indicate how two sub-diagrams are connected together.

In Fig. 1, some of the logical operations are very complex and others are very trivial. This is one of the reasons why the flow diagram is such an important tool. Any one of these logical operations may be replaced by a more detailed flow diagram of logical operations and we are still left with a flow diagram. In this way, it is possible to continually increase the detail in the diagram while still retaining the overall picture of the working of the program. This in fact is where the method used to write simple programs breaks down, the programmer cannot retain sufficient knowledge of all the paths through the program.

3 *Testing and running the equipment*

We have, in Fig. 1, a rough sketch of our program, let us look at it critically to see whether or not it will produce the information that we require. I consider that it has three quite serious shortcomings. There is no facility available for processing several pieces of paper tape. There is no way of varying the amount of monitoring information generated by the program. If an initial run shows up a defect on the paper tape record then there is no way of correcting or editing the tape even when we can see what went wrong.

At the development stage it is frequently useful to process a day's output or even less, just to ensure that certain equipment is working. At a later stage several of these short runs of tape may need to be combined. At the production stage it might be desirable to process the paper tape weekly and then at a later date combine four such tapes to give monthly figures. An alternative to handling several small tapes is to splice them together to produce one long tape. However, you may find that the computer installation looks unfavourably on this technique. The rollers on the tape reader squeeze the adhesive from between the paper tape and the backing of the adhesive tape. This adhesive then clogs up the tape braking mechanism and the area of the photoelectric cells. Even if the installation does not object to splicing paper tape a further difficulty arises. Paper tape is sold in 1000 ft reels and it may well be that a longer continuous record is required. Most computer installations are not even geared to handling full reels of paper tape and generally speaking a 200–400 ft run is much more convenient to handle. The problem is easily dealt with by giving the program the number of tapes to be read. The program then counts the number of tapes it has read completely. If it has not read all the tapes then it returns to reading the next one instead of printing out the results. The modifications to the flow diagram are shown in Fig. 2.

The program as it stands contains a simple choice, accumulate the data or reject it. If the data is rejected then it is printed out so that the reason for rejection may be ascertained. During both the development of the program and the development of the equipment this is an inadequate procedure. For both of these stages we require to know what information the program is receiving and what it is doing with it. At these stages it would be better if the program printed out all the information it received but we would then be swamped by output during production runs. The simplest way to deal with this problem is to impose two sets of bounds on each datum. If the datum falls between the inner bounds then it is

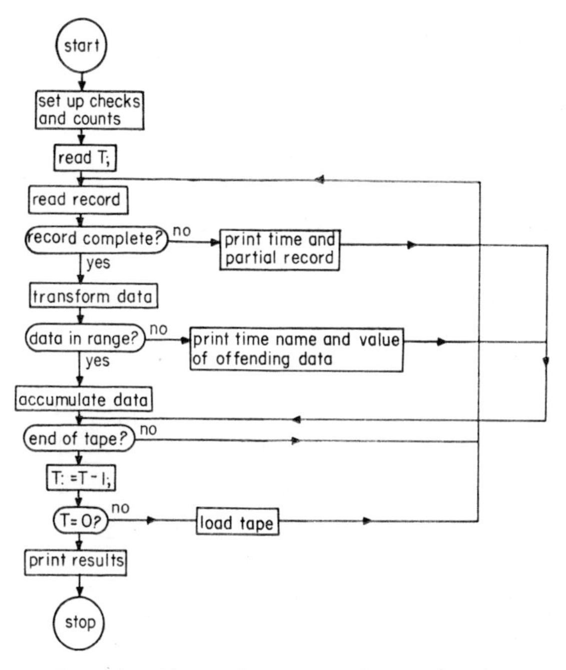

FIG. 2. Processing T tapes from automatic recording equipment.

only processed. If it falls between an inner bound and an outer bound then it is both processed and printed. As before, data falling outside the outer bounds is printed but not processed. The bounds are read as data at the beginning of the program. We may therefore vary the bounds from run to run and from variable to variable. We can, for instance, print out all carbon dioxide concentrations but no temperatures. It is therefore possible to obtain information from those parts of the equipment that are suspected of being faulty. The flow diagram, Fig. 3, is a modified version of that shown in Fig. 2. It represents a program which will process several tapes and also produce variable amounts of monitoring information.

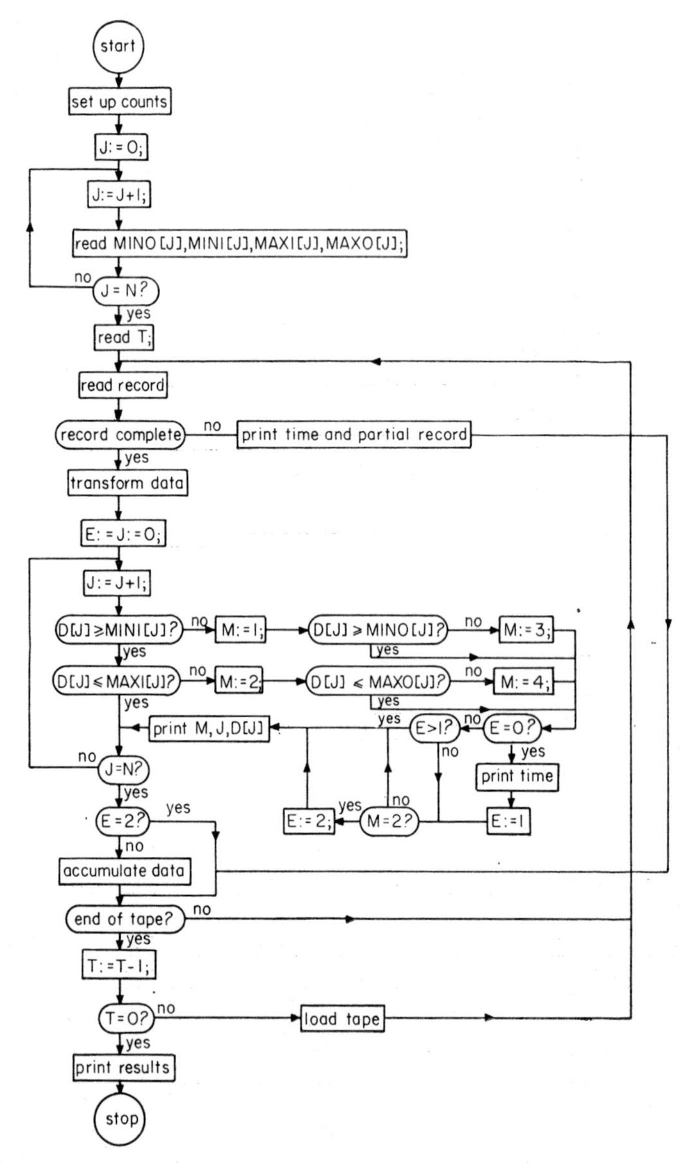

FIG. 3. Imposing a simple credibility check on each variable.

Editing facilities are very tedious to program but their value is such that it is invariably well worth the effort to include them. Consider the following problem, a paper tape with about a month's information on it has been produced. An analysis of the information for the second week only is required. If the program could obey the three instructions:

1 Ignore all records until the end of the first week.
2 Process all records until the end of the second week.
3 Ignore all records until the end of the fourth week.

then our problem is solved. If the program is incapable of accepting these instructions then the following procedure would have to be adopted. Unwind the tape and find the first and last records of the second week. Copy the segment of tape containing these records and process the copied tape. The copying is a potential source of further punching errors. This procedure is so tedious that it would seldom be performed. The problem described above is essentially that of masking parts of the paper tape. Another useful facility is that for correcting the odd punch error. There are two simple ways of doing this. The first processes the tape up to but not including the erroneous record and then replaces the erroneous record by information on the edit card. The second inserts an additional record immediately, this method is essential for amending the beginning of a tape. We have introduced four different editing instructions or commands:

1 Process tape up to and including the record at time TE.
2 Ignore tape up to and including the record at time TE.
3 Process tape up to time TE but replace record at time TE by the following information.
4 Insert the following information immediately.

Fig. 4 shows the necessary modifications in the flow diagram in order to allow these editing facilities. The program requires a further piece of initial information, the number of edit cards to be read NE. After the last edit command has been read then the program inserts a dummy edit command which causes the rest of the tape to be processed.

4 *Data transformation and accumulation*

So far we have ignored the problems of transforming data from one set of units to another and calculating means, maxima and other statistics. The reason for this is that compared with the problems of monitoring

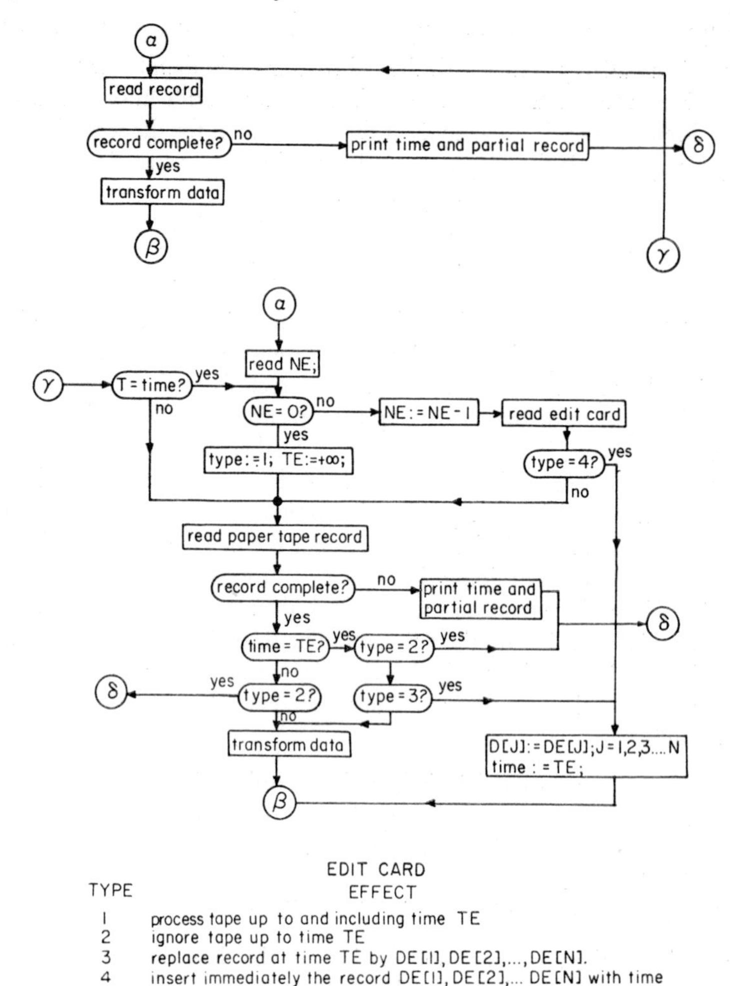

FIG. 4. Editing the output of automatic recording equipment.

and checking they are insignificant. If you wish to change the scale of an observed quantity or derive a quantity from those observed then in general you will have a formula for doing this. This formula can appear in the program in exactly the same form as it would be written on a sheet of paper. These formulae would appear in the operational box 'transform data'. Usually the computer will be able to transform and generate data much more cheaply than special equipment incorporated in your recording equipment.

The summation of data in order to compute means and standard errors is only a little more complicated. These operations are the type that are illustrated in the short programming courses.

5 *The use of larger computers*

The type of program that has been described above is designed specifically for small computers. That is a computer with no peripheral storage devices such as magnetic tapes, drum or discs and an insufficiently large core store to hold all the information on the paper tape record. A paper tape reader only reads in the forward direction and has to be reloaded to scan the tape a second time. It is therefore only convenient to process the tape sequentially. As we have mentioned, this procedure allows such statistics as means, maxima and minima to be computed. Histograms and frequency distributions can also be generated.

If the computer has peripheral storage facilities then multiple scanning and modification of the paper tape record can be performed. The entire tape is first read into the computer and stored. Punching errors will still cause the loss of records at this stage. The second stage is to scan the entire set of data for suspected errors. If a suspected error is found then both its predecessors and successors are available to the program and hence it may be replaced by an interpolated value. If the observed variables change slowly with time then this is an ideal way of dealing with isolated punch errors. Great care must be taken to detect faulty sensors. If a zero signal is sent in each record then no amount of arithmetic will be able to retrieve the correct or even estimate reasonable values for the variable in question. Having completed the error detecting and correction scan, the necessary statistics can be computed during the next scan.

Many modern computers have digital graph plotters. Such a device can be used to plot the trace of a complicated expression against time. This method of plotting will reduce the cost of the recording equipment by removing the need for the hardware to generate the complicated expression. In general, it is convenient to plot one graph at a time. This restricts the number of graphs that may be drawn on a small computer to one or two. On a large computer one graph can be produced during each scan of the data so there is no real limit. The digital plotter is a very versatile output medium which with suitable programming can be used to draw almost anything.

7 *Conclusions*

If you are contemplating setting up some data recording equipment then consult both the statistician and programming staff at the earliest opportunity. Their advice will almost certainly speed up the production of results and may result in a saving in cost or an increase in the information obtained. Whether or not you consider programming the computer yourself, a sound appreciation of the way a computer works and what jobs it does well will result in a better analysis of the data.

DEMONSTRATIONS

TEMPERATURE

Indicators and recorders for dry bulb temperature, wet and dry bulb temperature, and humidity. Range includes a very small clockwork temperature recorder and remote recorders. **Cambridge Instrument**

Thermographs. **Short and Mason**

Thermometers and thermographs including remote recording thermographs. **Darton**

Thermographs. **Casella**

Thermographs and thermistor thermometers. **Shandon Scientific**

Outfit for measuring rectal temperatures of small animals. **Shandon Scientific**

Battery operated thermistor thermometers and thermometer recorders. Up to nine points may be recorded continuously or at selected intervals. **Grant Instruments**

Battery operated thermistor probes for the exploration of large grain bulks. **R.W.Howe**

Battery operated device for measuring temperature using thermistors, and humidity using Humistors (a metal/polystyrene sensor). The output of an a.c. bridge is read from a meter after amplification and rectification. Output can be recorded on commercial portable miniature recorders. **P.T.Walker**

Silwood compensated thermistor bridge. Designed some years ago for temperature measurements in tropical stored products, the novel feature of the bridge is its ability to accept unmatched, and hence inexpensive, thermistors without the necessity to use a separate calibration curve for each thermistor. A second dial is set to the value of a constant peculiar to each thermistor, whose output then matches the calibrated temperature directly displayed on the main dial. **J.W.Siddorn**

An inexpensive thermistor probe developed to investigate the effects of temperature changes in the soil on gypsum resistance blocks.

The thermistors used show certain advantages over other types of electrical temperature transducers in soil temperature measurement. These include the following: (a) Resistance value within the range of bridges used to measure gypsum block resistances. (b) Good sensitivity (approx. 100 ohms resistance change/°C). Fairly simple portable field instruments can be used for measurements. (c) Good stability. Sample units tested over three years under different environmental conditions (a −5°C cold store, a forest drain, and on an exposed roof) have only shown a drift of the order of 0·5°C from the original calibration. However, cycling to high temperatures (40–50°C) has produced greater drifting.

Each unit consists of a Standard Telephones type K 13 thermistor and two resistors (Radiospares Ltd.) whose function it is to bring the temperature/resistance characteristic for each unit into close agreement with a standard. (Details of this procedure can be obtained in part from the Standard Telephones thermistor data booklet.) A lead of the required length is then attached and the assembly encapsulated in Araldite epoxy casting resin (MY753 and hardener HY951 from CIBA (A.R.L.) Ltd.). The calibration is checked at 0°C, 10°C and 20°C in constant temperature baths and the unit is then ready for use. Field readings are taken with an a.c. bridge.

W.H.Hinson, H.Gunston, D.F.Fourt

Thermistor leaf-clip and sensitive multi-range thermistor thermometer for measuring leaf temperatures. A simple leaf-clip holds a thermistor bead against the leaf lamina. Full details of this clip and of the circuits for measuring thermistor temperature to 0·01°C are described in Acock (1967). **B.Acock**

N.I.A.E. battery operated temperature integrator with digital read out (Weaving, 1967). **G.E.Bowman**

Huggins Mark IX Infrascope and calibration light source. For measurement of temperature without contact. **Claude Lyons Ltd**

See also pp. 266–7, 284–6.

RADIATION

Gunn-Bellani radiation integrator; the volume of liquid distilled by the radiation is measured. Kata thermometer. Black globe thermometer. **Baird and Tatlock**

Bimetallic actinograph (Blackwell, 1953). Sunshine recorders. **Casella**

Solarimeter and battery powered hourly integrator with digital output; this solarimeter is that suggested by Monteith (1959). **Lintronic**

Colour and cosine corrected light meter. Light meter for use amongst crops. **Megatron**

Battery operated integrating light meter with cosine correction, for wavelengths from 400–1100 nm; silicon photovoltaic diodes produce an electrical output proportional to the incident light. **Plessey**

True cosine reponse spectroradiometer by ISCO for wavelengths from 380–1050 nm. A planar photodiode is used with a battery or mains operated transistor circuit. Bandwidth is 15 to 30 nm. Programmed scanner and recorder for this instrument. **Shandon Scientific**

Radiometer by YSI utilizing small dual thermistor bolometer probe allowing use from 250 to 3300 nm. Seven ranges from 0·25 to 250 mw/cm^2 full scale. **Shandon Scientific**

Silwood submersible light meter. Silicon solar cell detectors are mounted at the centre of an integrating sphere made from a table tennis ball (Powell and Heath 1964). Included within the sphere are colour filters to counteract the bias in spectral sensitivity of the solar cell, a mask to match the two sides of the detector, and silica gel to absorb any moisture diffusing into the head. The head is mounted flexibly, to avoid accidental damage, at the end of a telescopic aluminium handle. Calibration is arbitrary, since the instrument was designed for comparative readings of light intensity in muddy ponds containing water snails. **J.W.Siddorn**

Silwood actinometer. A spherical thermopile, mounted in a 'Pyrex' glass flask, integrates short-wave solar radiation over 360° of solid angle It is used in the field to sum direct, diffuse and reflected thermal short-wave radiation. **J.W.Siddorn**

Thermopile solarimeters of tube and dome type for field use (Monteith 1959; Monteith and Szeicz 1962; Szeicz, Monteith, and Dos Santos 1964; Szeicz 1965). **J.L.Monteith**

A system for the measurement and recording of radiation penetrating leaf canopies of field crops. Modifications of the Szeicz, Monteith & Dos Santos (1964) tube solarimeters were made as follows: Solarimeters of six elements each 12 cm long were mounted in 30 in. long Pyrex tubes containing silica gel. Each element comprised 240 copper-constantan junction pairs painted black and white. Alternate sections were wound with opposite phasing. Three of the elements in each tube were connected in series and shielded from radiation below 735 nm with Wratten 88A gelatine filter. The other three were also series connected and the two sets connected positive to positive. The potential across the negative ends of the six elements was then proportional to the visible radiation. Reference solarimeters were calibrated against the Kipp at the N.I.A.E., Silsoe and was approximately

150 mV/cal 400–735 nm/cm²/min. The solarimeters in the experimental plots were calibrated *in situ* over a 2-day period prior to plant emergence. A similar recalibration was done after all plots had been harvested. Relative sensitivity of solarimeters beneath the crop decreased with time as they became dirty.

In the field the solarimeter tubes were arranged radially in threes and wired in parallel. They were mounted on 8 cm cork rings and levelled.

The output from 48 such sets of solarimeters was fed in sequence at 30-second intervals via a sequential switch to a Kent Multilec III pen recorder. The reference voltage for the slide wire was taken from a control set of solarimeters mounted above the crop suitably matched via a small d.c. amplifier with a gain of 1·4. The recorder thus read the proportion of radiation reaching the ground, with a scaling factor of 1·4 to allow for the possibility of readings in excess of the reference.

Twice in each 30-minute sequence the reference solarimeters were read directly in mV, as were a similar set mounted facing downwards to measure reflected radiation. Instructions to switch from relative to absolute readings were initiated from a program drum attached to the 60 channel sequential reed selector switch.

A time reference was punched before each 30-minute sequence using a digital clock stepping from 1 to 48 at half-hourly intervals.

The pen recorder slidewire drive shaft was coupled to a Benson Lehner rotary digitizer, the output of which was fed to a data logger. The digital information was decoded, then re-encoded in IBM code and punched on to paper tape on receipt of punch command instructions from the program drum. During the punch sequence the slidewire servo motor was disconnected to prevent movement of the digitizer.

Paper tape was converted to punched cards for ease in editing, and typewritten output obtained. Corrected and edited cards were then used to calculate the per cent light absorption by the crop canopy for each treatment (mean of three plots) at 30-minute intervals throughout the day. A daily summary was then printed, together with integrated values of total incident, reflected and absorbed radiation. Temperature means for day, night, and total were calculated.

D.A.Green, L.H.Jones & N.J.T.Melican

Integrating photometer by National Institute for Agricultural Engineering.

G.E.Bowman

Electrolytic solarimeters. Powell & Heath (1964) described a cheap and simple solarimeter based on a selenium photocell and a copper voltameter, which they used in glasshouses. Similar solarimeters have been used to determine the light reaching various parts of a reedswamp in comparison with the light reaching an open site during the last nine months. Various practical problems have arisen from leakage and corrosion but can be overcome. Considerable and inconsistent differences between cathode gain and anode loss have been noticed possibly because of the formation of copper oxide and some loss of deposit while returning the electrodes to the laboratory. Macfadyen (1956) used silver voltameters containing 0·25 per cent acetic acid in a 25 per cent silver nitrate solution for temperature integration. Such silver voltameters in series integrated currents of 5–250 μa accurately over a week (for 5 μa the 5 per cent confidence limits were 2 per cent of the mean) and a voltameter and photocell combination in the field has performed satisfactorily. **D.F.Westlake**

Electrolytic solarimeter based on the design of Powell and Heath (1964). Filters and light diffuser over the selenium photocell give approximately equal response to light over the range 400–640 nm at any angle above the horizontal. The integrator is separate and the whole is very robust. The current is integrated over periods of a week or a fortnight by determining the weight of copper transferred between the electrodes. **J.E.Jackson & C.H.W.Slater**

Hemispherical photography in studies of plants (Coombe & Evans 1966). Also exhibited was an overlay grid in which each segment corresponds to 0·1 per cent of the total illuminance from a standard overcast sky (Anderson, 1964). **D.E.Coombe & G.C.Evans**

Apparatus for surveying light climate in woodlands, including the area and intensity of sunflecks (Evans 1966*b*). **G.C.Evans**

Integrating sphere for spectral measurements on leaves. In ecological contexts the value of radiation measurements in absolute terms is often much reduced because insufficient information is available on the spectral composition of the light and on the spectral response of the particular ecological process being studied. Such information is difficult to obtain with standard equipment, because most commercial spectrophotometers and detectors are designed for use with parallel light and are unsuitable for use in natural environments in which light is spatially distributed through at least a whole hemisphere.

This is a simple and relatively cheap device for determining the spectral transmission and reflection of a leaf. It can be used for two purposes: 1. To characterize the spectral composition of the light climate under a continuous canopy of similar leaves; 2. To estimate the heating effect of light of a given spectral composition falling on a leaf. This is a major term in the energy balance.

Summary of procedure. An intense parallel beam of white light is passed through one of a series of narrow-band interference filters. The resulting approximately monochromatic beam falls either on the specimen or on a standard white reflecting surface.

The diffuse light reflected or transmitted is collected by an integrating sphere. Integration depends (French 1960) on the fact that, when light passes through a small hole in a sphere painted a matt white inside, the intensity and directional composition of the light emerging from a second small hole depend only on the *intensity* of the incoming light and not on its directional composition. If a photocell is placed at the second hole, therefore, it records a fixed fraction of the light coming from the first hole (at a given spectral composition) regardless of its directional composition. An essential condition is that the holes should be screened from one another by a white disk so placed that the photocell cannot see the entrance hole but only the walls of the sphere.

In order to measure reflection and transmission with the same sphere, it is provided with three holes, two polar and one equatorial (Fig. 1). The equatorial hole is used for the photocell. The light beam is aligned with the other two holes, which will be referred to as the entrance and exit holes. To measure *reflection* (Fig. 2), the entrance hole is left vacant, and successive readings of the photocell are taken (a) with the leaf at the exit hole, (b) with the standard surface substituted for the leaf, (c) with nothing at the exit hole (in order to correct for scattered light resulting from imperfect collimation of the beam). To measure *transmission*, the standard white surface is placed at the exit hole, and readings are taken (d) with the leaf at the entrance hole and (e) with nothing at the entrance hole.

Since the entire leaf is outside the sphere, it can remain attached to a branch or to a potted plant. There is thus no difficulty in keeping it alive during the experiment.

Determinations should be made in a dark-room to avoid stray light.

The magnitude of any one reading depends in a complex way on the spectral properties of the light source, the sphere paint, and the photocell, as well as of the leaf. Since one is comparing a reading with the standard at the same wavelength, however, this dependence does not need to be known in order to obtain an absolute value of transmission or reflection.

A disadvantage of this method is that the incident beam is parallel and usually impinges at normal incidence. The spectral properties of a leaf may vary with the angle of incidence. Such variation (at least

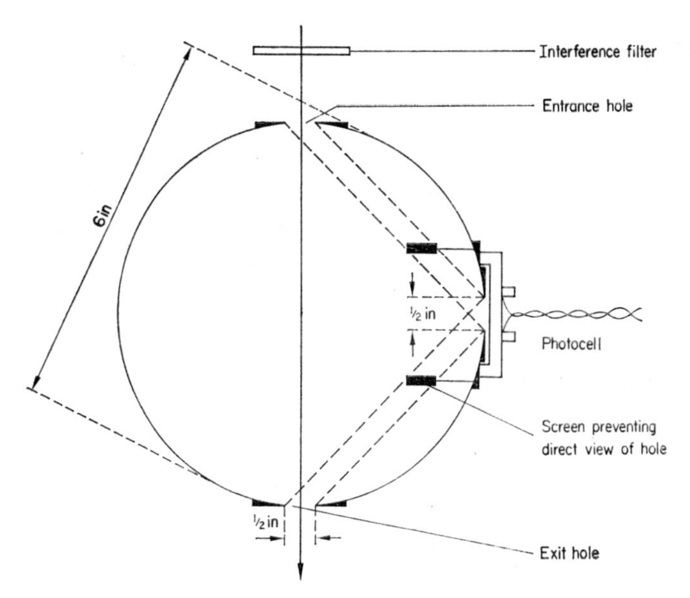

FIG. 1. The integrating (sphere. 1 in = 2·54 cm).

as regards transmission) could, of course, easily be studied with this apparatus. The exhibitors are interested mainly in the light climate of broad-leaved plants on the floor of a wood. Many of these plants have horizontal leaves. Most of the incident short-wave radiation comes from near the zenith (Evans 1966*a*). Hence the requirement of normal incidence is approximately fulfilled in this case.

Details of Components. The light source is a quartz-iodine lamp of the type used in projectors. It is a small very intense source with good emission in the blue. It is usually necessary to run it from a good voltage stabilizer or a large series of accumulators.

A series of interference filters (11 is an adequate number for many purposes) is preferred to a prism or grating monochromator for the following reasons: (1) It gives an intensity sufficiently high (even after absorption in the later stages of the system) for a photocell, rather than a photomultiplier, to be used as a detector. (2) Prism or grating systems transmit stray light in regions of the spectrum remote from that being studied. This would be a very serious disadvantage

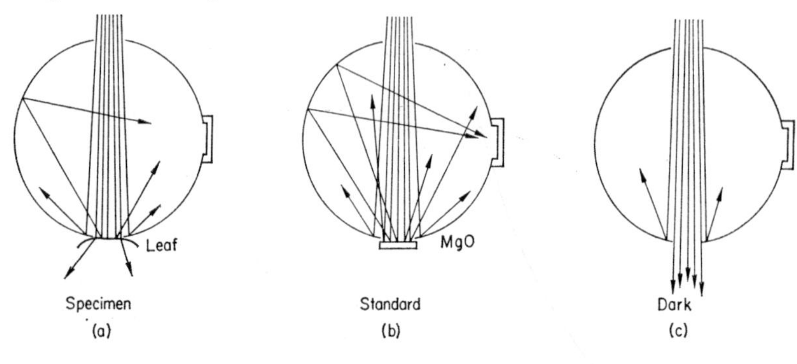

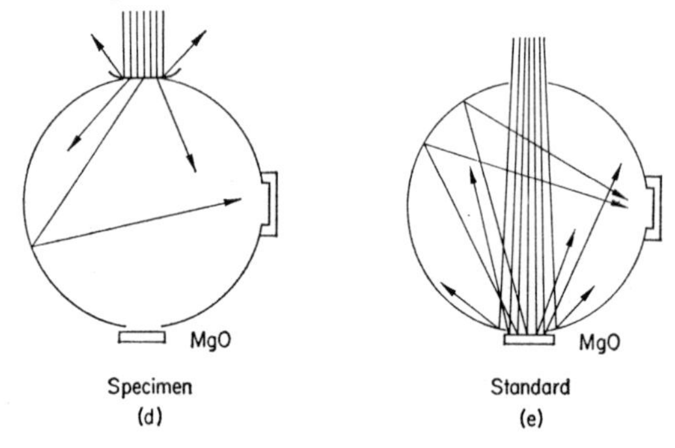

FIG. 2. The integrating sphere showing stages in determining reflection and transmission.

for the present application. A double monochromator removes stray light but at the expense of an even weaker emergent beam. The 'tails' of unwanted wavelengths transmitted by interference filters are much less serious because they are close to the desired wavelength.

The diffusely-reflecting reference surface is of magnesium oxide deposited on the end of a metal piston sliding in a tube with a broadly-flanged end. In use, the piston comes almost to the end of the tube so that the oxide surface projects slightly beyond the flange. When not in use, the piston can be retracted and the whole component stood upright on the flange. This protects the oxide surface from dust and mechanical damage. A further advantage is that the oxide can be deposited with the piston in the forward position to avoid the difficulty of getting an even deposit on a small surface.

The detector, a Megatron Infra-red photocell, gives a reasonable output in the wavelength range 450–1000 nm. The irradiation of the cell is so low that there is no difficulty in securing a response that is linear with intensity. The sensitivity of the recording instrument must be at least 1 μV; a sensitive Pye galvanometer is adequate but an amplifying voltmeter is an advantage.

We are much indebted to Dr B.H.Crawford, of the National Physical Laboratory, for most valuable discussions, and have made use of information supplied by the N.P.L. on magnesium oxide smoking and other technical matters. **O.Rackham & Joan Wilson**

Point quadrat apparatus. This apparatus was originally designed to increase the speed and accuracy of point quadrat measurements on 160 cm square plots of dense vegetation. Errors arose due both to difficulty in observing point contact (especially in dense heather and *Molinia* tussocks), and to manual lowering of the probe allowing an operator bias. The apparatus designed to overcome this operates on dry batteries and consists of: (a) An electronic contact indicator. (b) An electrically controlled system to move and record the position of the probe. The recording system also has a code to indicate such things as the part and species of the plant in contact with the probe tip.

The electronic system designed by Mr D.T.Smith (Clarendon Laboratory, Oxford) operates on a 12-volt dry battery and consists of an electrometer valve (Mullard ME 1401), microammeter and oscillator. The valve grid is normally negatively biased and in this condition no current flows to the microammeter. The grid is connected to the probe tip and the positive terminal of the battery is connected to earth, when the tip makes contact with vegetation the grid becomes positively biased and the valve passes current. This is indicated both visually by the microammeter and audibly by a change in the frequency of the oscillator. Unfortunately the instrument will not detect contact with dry dead vegetation because of its extremely good insulating properties, however this may be overcome by damping the vegetation with a light spray of water.

The tubular aluminium point quadrat probe is located on a plate by four guide and two pressure wheels. One pressure wheel is connected to a 24-volt motor which is remotely controlled to raise or lower the probe. The second pressure wheel is linked to a low torque rotary potentiometer which is connected via a control box to a 'Limpet Logger' (d-mac Ltd.) One rotation of the potentiometer is equal to a probe movement of 100 cm.

The operator uses a control keyboard to trigger the 'logger' whenever the probe tip makes contact with vegetation and by using different trigger buttons records separately a code for the part of the plant or species, besides the position of the probe. For grasses a jet of air from the probe tip is successful in clearing the leaf after each contact to prevent it fouling the probe when it is lowered further into the vegetation.

The whole system is mounted on a trolley and can be angled for inclined point qadrats. The trolley runs on a 'Dexion' gantry which in turn runs on a frame so that random points can be sampled over an area. **F.B.Thompson**

Point quadrat apparatus for measurement of foliage distribution and light interception. Records of the number of contacts made with foliage by the point of a needle as it is gradually lowered through a canopy at specified angles can be used to estimate the foliage area per unit area of ground and the inclination of the foliage to the horizontal. These characteristics are important determinants of light interception.

The line along which the needle descends is comparable to the path of a ray of light, and the first contact with foliage is comparable to the point of interception of the light ray. Consequently, point quadrats inclined at a particular angle give information on interception of light at the same inclination (WarrenWilson 1965).

J.Warren Wilson

A small direct calorimeter at present under construction at the Rowett Research Institute, Aberdeen was shown together with typical heat loss records. **J.D.Pullar & J.M.Brockway**

Artificial sheep, developed at the Hannah Dairy Research Institute, Ayr, mainly by K.L.Blaxter and J.P.Joyce. A cylindrical container roughly the size of a sheep is maintained at normal sheep body temperature by thermostatted internal heaters. The power consumption of these heaters gives a measure of the environmental demand, and makes it possible to compare and evaluate different conditions of temperature, wind speed, rain and solar radiation in terms of the sensible heat loss of the artificial animal. **J.D.Pullar & J.M.Brockway**

WIND SPEED AND DIRECTION

Remote indicating wind speed and wind speed and direction indicators. **Short & Mason**

Portable vane operated air meters. Recording manometer for 0–6 in. water gauge. **Darton**

Cup counter, cup contact and cup generator anemometers. Wind vanes including remote indicating wind vanes. **Casella**

Self-balancing low speed anemometer with a range of 0–1·7 m/sec and which is most sensitive at the lowest wind speeds. It is adapted from the design of Head & Surrey (1965) and Head & Thorp (1965). **J.S.Kennedy**

A hot wire anemometer for the accurate measurement of wind velocity and turbulence. A platinum wire 0·01 mm in diameter and 1 mm in length is heated and maintained at a constant temperature of the order of 200°C. The current required to maintain a constant temperature is a measure of wind velocity. The very fast response time of the sensor (5×10^{-5} sec) means that high frequency fluctuations in wind velocity are detectable, providing a turbulence signal suitable for further analysis. The anemometer provides, as alternative outputs, wind velocity and turbulence intensity.

This type of anemometer has for many years been a tool of aerodynamicists. Its successful application to environmental measurements, however, is recent and requires special precautions owing to the fragility of the sensor. The small size and sensitivity of the sensor make it a useful instrument for measurements in leaf canopies, or in the very thin boundary layers of leaves or animals in connection with heat transfer studies. See Somerville & Turnbull (1963) and N.P.L. notes on applied science No. 33. **G.B.K.Baines**

A simple heated thermocouple anemometer. Five pairs of copper-constantan thermocouples inserted in a tube of 5 mm diameter and 7 cm length are so arranged that one of each pair is heated by a manganin heating wire coiled about the upper end of the tube while the second, in the lower end of the tube, remains at ambient temperature. The millivolt output of these thermocouples, connected in series, varies inversely as the wind velocity. With this arrangement a heating current of 0·5 A produces, in still air, an output of approximately 10 mV.

The small probe size makes this an ideal sensor for measurements among leaves. It is robust, simply constructed, and very sensitive in the low velocity range up to 1 m/sec encountered inside plant canopies. **G.B.K. Baines**

Bidirectional wind vane. This vane provides a simple means of assessing the relative turbulence in the wind above a rough surface. The response of the vane to eddies in two dimensions over a short period of time is recorded on a chart in the form of an ellipse whose vertical and horizontal dimensions are related to the dimensions of the eddies passing during the period of exposure. The mechanical properties of the instrument determine the range of eddy sizes or frequencies recorded. Instruments of three sizes are in use in Bramshill forest, Hampshire. See Taylor (1927) and Thompson (1953). **E.R.C.Reynolds**

Environmental winds at levels well above the ground, for example those encountered by migrating birds and insects, can be measured by means of a pilot balloon filled with hydrogen to ascend at a known rate and tracked by theodolite. From measurements at regular intervals of elevation, azimuth and altitude, the position of the balloon can be calculated and the mean wind speed obtained in successively higher layers in the atmosphere. **R.C.Rainey**

HUMIDITY

Hair hygrograph. Hair hygrothermograph. Humidity indicator. **Short & Mason**

Animal membrane hygrographs and hygrothermographs by Bacharach Industrial Instrument Co. **Shandon Scientific**

Humidity indicators. Hair hygrographs and thermohygrographs including distant reading model. **Darton**

Indicators and recorders for wet and dry bulb temperatures, including distant recording model. **Cambridge Instrument**

Assmann hygrometer and other wet and dry bulb hygrometers. Hair hygrographs and thermohygrographs. Humidity slide rules. Hygrometric tables. Dew point apparatus. **Casella**

Battery operated thermistor psychrometer indicators and recorders. Up to nine channels can be recorded continuously or at intervals of up to one hour **Grant Instruments**

Relative humidity and temperature recorder and recording controller; mains operated strip chart recorder using aspirated wet and dry bulb sensors. **Foster Instruments**

Portable recording psychrometer mast. A portable battery-operated recorder with eight channels (Grant Instruments) is used to record the signals from eight thermistor thermometers. These are mounted in four pairs above and within a crop canopy in simple unaspirated psychrometer heads which are based on a Rothamsted design. The resulting data can be used as the basis for calculating the Bowen ratio, which gives a measure of the flux of water vapour over the canopy and therefore an estimate of evaporation rates. Full details in Borhan & Acock (1967). **A.Borhan & B.Acock**

Low-cost thermocouple and thermistor hygrometers developed for measuring relative humidity in air lines. Humidity control unit. **J.R.Etherington**

RAINFALL AND EVAPORATION

Rain gauges. Tilting bucket rain gauge for recording on site or at a distance. **Darton**

Rain gauges. Natural siphon and tilting siphon rainfall recorders. Jardi rate-of-rainfall recorder. Electrical rainfall recorder for mains or battery. **Casella**

Plot runoff collecting and measuring apparatus. Runoff is collected, via a filter and settling pot (to remove entrapped air) in standard 55-litre, 22-gauge drums (14×25 DL Kegs, Drums Ltd.). These can hold rainfall equal to 55 mm of rain on the square metre plots and are emptied at weekly intervals by means of a tap (except in one case where an automatic siphon system is used). The water level in the drums is sensed by a counter-weighted, 30 cm diameter expanded polyestrene float which actuates a low torque (7·24 gm/cm) 360° precision potentiometer by means of a glass fibre thread and winding drum. The smallest amount of rainfall to be recorded, 0·05 mm, can cause a torque change of about 35 gm/cm in the system which requires a torque of 14–21 gm/cm to operate.

Intensities of up to 99 mm/hr are allowed for by recording the float position every three minutes by means of the potentiometer and a 'Limpet Logger' (d-mac Ltd.). In case this intensity is exceeded at least one drum per site is designed with a sensitivity of 0·10 mm so that it will record intensities up to 198 mm/hr. The float position can also be read by means of a numbered wheel and colour code system connected to the potentiometer shaft. **F.B.Thompson**

Stem flow on trees. It is desirable to collect stem flow for measurement or analysis without damaging the tree trunk. Deformation of the stem growth if the gutter is left in place for prolonged periods is also undesirable. Aluminium coach guttering wound around the stem is reasonably successful in achieving these objectives. Irregularities of the bark are shaved off or filled with Plastic Putty (Kelseal, type 400). The guttering is hammered gently to the shape of the trunk and its upper edge painted at frequent intervals with 'Synthaprufe' (National Coal Board, Cardiff) as the tree expands the spiral. Precautions are taken to prevent water running along the underneath of the spiral. The bottom end of the gutter is conveniently bent into a cylinder for insertion into PVC tubing. **E.R.C.Reynolds**

Leaf wetness recorder. The instrument is based on changes in resistance between electrodes due to water on an intervening leaf surface.

A dispersion of flake silver in methyl isobutyl ketone is used as the electrode material painted onto 'melinex' film. Gaps in the film allow it to be threaded onto a leaf so that the electrodes are separated by the leaf surface (Fig. 1, p. 106).

The electronic circuit, designed by Mr D.T.Smith (Clarendon Laboratory, Oxford) uses transistors and can operate for long periods in the field using a 12-volt dry battery. It consists of an oscillator which feeds a Wheatstone Bridge circuit in one arm of which are the electrodes; the maximum voltage applied to the electrodes is about 3 volts. The bridge output is fed to an a.c. amplifier which in turn feeds a phase sensitive detector (necessary to show which way the bridge is off balance). The signal from this is fed either to a recording voltmeter ('Limpet Logger' by d-mac Ltd.) or to a switching transistor which operates an event recorder via a relay. The balance point of the bridge and hence the output characteristics of the amplifier in relation to the inter electrode resistance can be varied by means of a one megohm variable resistance on another arm of the bridge.

F.B.Thompson

Surface wetness recorders. Since 1958 there has been considerable effort to develop a reliable wetness recorder for fungal disease studies. In the British model an expanded polystyrene surface is used to simulate a leaf surface, and the deposit of rain or dew is, in effect, weighed, the changes in weight being recorded on a drum with a 24-hr chart.

Recent work with the Meteorological Office has shown, that for certain purposes, nearly equivalent information can be obtained by utilizing standard meteorological data, combining hours of duration of rain with hours of high humidity. It might be that this has wider usage, for example in work with Bryophytes, Pteridophytes, Lichens and their ecology. (Hirst, 1957; Preece & Smith, 1961; Preece, 1964; Post, Allison, Burckhardt & Preece, 1963, this last contains the only available review of other types of surface wetness recorders.)

T.F.Preece

Meteorological Office leaf wetness recorder (MK 3A. Ref. No. 2001). Designed originally for potato blight and apple scab warning, the weight of the polystyrene foam block is recorded on a daily strip chart. The earliest design employed a weighed living shoot supplied with water through its cut end. The simulated surface is subject to weathering (compare fresh and used blocks). The effect of wind is reduced by a paddle in silicone damping fluid. Records show that even dew formation on the polystrene surface is recorded. (Hirst, 1957; Hirst, Long & Penman, 1954). **E.R.C.Reynolds**

Field evapo-transpirometer. The apparatus measures the water balance of a block of substrate (peat) and its associated vegetation. It is designed so that the limits of the range of fluctuations of the water table are controlled so that the overflow can be measured. The input of water as precipitation is measured by an adjacent rain gauge. The performance of the vegetation or of individual species in the container may be measured and related to the water regime. These water regime data may be used to estimate those for a whole ecosystem, e.g. a peat mire. **M.C.Pearson**

Drainage lysimeter operated by remote control. This is a cheap and robust method of measuring the balance between rainfall and evaporation from a plant canopy, in which all operations are carried out from a distance of a few yards to avoid disturbing the plant cover. By regular irrigation, again by remote control, the apparatus can be used to measure evaporation (from soil and plant) under conditions where the actual rate of water loss is near to the potential rate. This pattern of lysimeter has functioned well in the open and under glass, under arid tropical conditions and in humid temperate climates. Details of construction and operation are given by Hudson (1967).

J.P.Hudson

A weighing lysimeter with load-cells. Weighing lysimeters are one of numerous methods that can be employed to find the evaporative flux at the surface, provided a few simple precautions are observed in the lysimeter design. The lysimeter requires: (1) A large surface and an important mass of soil (particularly for AET measurement). (2) No perturbation of water, vertical and horizontal circulation, thermal flux in the neighbouring soil, and aerial environment. (3) High sensibility because very low changes in weight must be recorded out of an important mass (0·5 kg/5800 kg for PET and 1·5 kg/14 500 kg for AET).

The lysimeter consists of a body of soil encased in an iron tank the sizes of which are:

	Diameter (m)	Depth (m)	Weight (kg)	Surface (m^2)
PET	2·45	0·60	5 800	4·8
AET	2·11	2·00	14 500	3·5

In the bottom of the lysimeter a drainage chamber, made with a drilled portion of a sphere, is covered with a bed of gravel (0·15 m high) and fine sand (0·10 m high). The soil is placed above the sand. A vertical furrow placed in the drainage chamber is provided for pumping out excess water.

The soil mass rests upon a weighing frame in two parts. The upper frame rests upon three load-cells of unitary capacity of 2000 kg (PET) and 5000 kg (AET). It has a steel ball to centre the tank of soil. The three load-cells are fixed on the lower frame which is centred on the bottom of the outer tank by a centering stop. The two parts of the frame are removable. Inner and outer tanks are joined by means of a flexible butyl rubber gasket. The space between the tanks is about 5 cm.

The signals from the three load-cells in one lysimeter are connected to a strip chart recorder (an automatic balancing electronical potentiometer model 'Extensometer' licence ONERA). The purpose being the recording of changes $\pm p$ on both sides of the weight P; the potentiometer involves a zero adjustment.

The sensibility of load-cells is better than 1/10 000, the accuracy of the strip chart recorder is 0·2 per cent of the range scale. This provides the following accuracy for the lysimeter:

	Range scale (kg)	sensitivity (kg)	accuracy on the chart (kg)	accuracy on the chart (mm water)
PET	− 100 to + 100	0·5	0·4	0·1
AET	− 212·5 to + 212·5	1·5	1·5	0·43

The advantages of this method of measurement are that: (1) Static sensors eliminate movable mechanical devices; load-cells accept for a while overloads of as much as 50 per cent of unitary capacity. (2)The characteristics of the load-cells eliminate lateral disturbing strain (essentially the wind effect). (3) Changes in temperature have no effect on the weighing system. (4) The measurement is independent of the arrangement of the strain. (5) The strip chart recorder can be several kilometers distant from the lysimeter (0 to 15 km) without effect on the accuracy. **Ph.Grebet**

An electrically-weighed lysimeter based on the use of strain gauges. Electric strain gauges are used to measure changes in weight of containers of soil, in which a large part of the weight is taken by strong helical springs. The method, which has shown considerable promise, has the advantage that the expensive parts of the equipment can be shared between any number of containers. In this apparatus the method is being used to follow changes of weight in 15 contianers, each about 60 cm square, and holding about 45 kg of soil in which crop canopies of various sorts have been grown. Further details in Hand (1967) **D.W.Hand**

Water level recorder. The instrument allows water levels to be measured accurately (± 1 mm) in peatland communities without the observer disturbing the water-table or damaging the vegetation immediately adjacent to the site. It consists of a movable vertical scale supported at one end of a horizontal aluminium tube 2 m long. The scale can be made to descend into a tube previously inserted into the peat. When the zero point touches the water surface an electric circuit is completed which switches on a light.

By this means it is possible for the observer to remain 1·5 m away from the site being measured. **D.A.Goode**

Cheap hydraulic lysimeter. The soil container consists of a large polythene domestic waste bin, and the outer container is a 180 l oil drum. Cost about £4. Accuracy about plus or minus 0·25 cm. The system is temperature sensitive; to deal with this, one replicate in each group is permanently sealed against water loss and its daily reading is taken as the zero for neighbouring crop-carrying lysimeters. **E.J.Winter**

Surplus/deficit indicators. Mk. III is a device for field use, intended to imitate the plant/soil system in respect of water uptake and loss. Dimensions are such that evaporation from the pad approximately equals potential transpiration from an area of foliage equal to that of the receiving funnel. The instrument is set up in the field in springtime and then the deficit built up in the container and indicated on a sight tube approximately equals the current soil moisture deficit provided that the soil moisture deficit is not allowed to become large enough to reduce transpiration. Accuracy is about plus or minus half an inch of deficit.

If a collecting vessel is attached to the overflow, the amount collected will correspond roughly with, for example, winter percolation and drainage.

In Mk. IV the ceramic evaporating surface (grooved to reduce rain outsplash, is formed from the ends of a series of vertical porous plates of different lengths separated from one another by a series of impervious membranes. As the level of water in the container falls (in harmony with increasing soil moisture deficit) the lower ends of the plates are successively left above waterlevel, and the area of the evaporating surface is reduced. The lengths of the plates are chosen so that the overall pattern of evaporation imitates the moisture release characteristic of the surrounding soil. **E.J.Winter**

SOIL WATER

An instrument for the determination of soil moisture content. Soil moisture determinations can be made by measuring the degree of heating of a coil when a known current is applied. The greater the moisture content of the soil the faster will the heat be dissipated thus preventing the heating of the coil to the value achieved in dry soil.

A specially shaped soil moisture probe containing the heating coil and a sensing thermocouple is placed in the soil. The heating current is applied for 10 sec from a 4·5 V battery and the thermocouple current is measured by a mirror galvanometer. Absolute values of soil moisture content can be estimated if the probe has been calibrated with the soil under investigaton. Different degrees of soil moisture in the field can be detected without calibration.

Reasonably uniform contact between the probe and the soil can be obtained with some practice. However, in such a heterogeneous system as soil the probe may accidentally come to be in a cavity, next to a stone or in a region of different compaction. Erroneous readings produced in these ways are easily detected since several readings can be taken speedily in any one locality. **H.Meidner**

A miniature gypsum resistance block for measuring soil moisture tensions. The block is designed for use in shallow soils or in any soil in which steep gradients of soil moisture tension are found. The overall dimensions of the block are $2{\cdot}5 \times 1{\cdot}25 \times 0{\cdot}6$ cm. Parallel electrodes are formed by removing a section of the insulation of twin core, multistrand cable, teasing the strands apart and flattening the two cores so that they form two parallel 'plates' 1 mm apart. The electrodes are then encased with plaster of paris which is impregnated with nylon, an extension of the cable forms the lead. The blocks are as satisfactory in their response as other gypsum designs and are simple and quick to make. **P.S.Lloyd**

Neutron moisture meter with equipment mounted on a wheel barrow.

W.H.Hinson, H.Gunston & D.F.Fourt

Gypsum resistance blocks. Our design of block differs only in detail from generally published descriptions. These details are: (1) The use of silver gauze for the electrode material makes soldering much easier compared to nickel or stainless steel. Silver resists corrosion and is an excellent conductor. (2) The electrodes are painted with Aquadag, a water suspension of colloidal graphite, before embedding in plaster. The effect is to reduce the electrolytic capacitance component. (3) To facilitate emplacement down augur-holes a cylindrical casting mould is used made of 500 gauge polythene. This reduces cracking during setting, allows rapid pouring and facilitates removal of the casting from the mould. (4) Mixing and casting are standardized procedures. 25 gm of dental grade plaster is added to 30 ml of water in a 50 ml beaker, stirred rapidly for 20 sec with a piece of 12 mm dowelling and poured into the mould. Bubbles are dislodged with a micro-spatula and several hours allowed for setting.

It is considered advisable to locate and prepare the site at a time when the whole profile is moist, and when recent rain has exceeded evaporation losses. The face of a pit or borehole can then be examined for seepages, as any impedance of percolation should be known before the work commences. Before emplacement the blocks should be completely wetted in distilled water, from the base upwards, and kept wet. There are two main emplacement techniques: (1) The pit method. Emplacement at the end of horizontal bores ensures that water reaches them through undisturbed soil. (2) The inclined-bore method. On favourable sites this technique can combine speed with minimum disturbance and provided the bores are very carefully back-packed, they will not provide lines of easy access for rain-water.

For the pit method of emplacement a large pit is excavated, 2 m long × 0·5 m broad, and about 2·2 m deep, oriented so that the stacks of blocks can be placed 1 m from each end of the pit. The site of a thinned tree-stem, or the centre of a group is a suitable place in the forest. The spoil is stacked on sacking or polythene sheets, by horizons, away from the pit sides. The pit should be kept to the stated dimensions with the ends vertical. The block positions are marked out on the end walls, and using a 1 m horizontal screw augur a hole is made of 'clearance' size for the block. A short 'tolerance' bore is then made at the far end of this hole. Using a tubular aluminium holder with a rubber thrust-pad, and with the block held by the cable threaded through the holder, the block is pushed into its 'tolerance' bore, and the holder removed. Water is used as a lubricant, and the bores are carefully back-packed with the material taken out using a long spoon and tamper. Thermistors can also be emplaced in these holes. The pit should be carefully back-filled and firmed about 30 cm at a time. The pit method offers excellent opportunities for obtaining cores, and other samples for laboratory calibration of the material, and a close-up examination can be conducted of root distribution, seepages, worm tunnels, fissures, and compact layers.

The inclined bore method of emplacement combines speed and ease of working with minimum of disturbance at the cost of more special tools and the need for extra care at certain points. It is not easy to thoroughly examine the profile. Inclined screw augur holes are made down to the required levels and, after emplacement as in the pit method, are backpacked to at least the previous density. Blocks from 15 cm to about 75 cm can be emplaced from near the surface with angles of about 45°. Deeper levels, to 2·4 m or so, are more easily reached by using the screw augur in the floor of a shallow pit penetrating the looser surface layers and using an angle of 70–75°. A large set of augurs from 1·2–3 m in length, and special holders and tampers have been found useful.

Some advantages of gypsum blocks are: (1) They are cheap, easy to emplace in large numbers, and free from maintenance troubles for at least three years. Some have gone five years and are still good. (2) Readings are rapid and easy, and unskilled personnel can be taught to carry out the routine. (3) Emplacement of these units in a large mass of soil such as a tree root-space should only cause a very small disturbance of the environment if back-packing and site detail are of a high standard. (4) Measurements reflect soil tensions and are useful for identifying tension gradients. (5) Although relatively stonefree soils are easiest, with patience blocks can be emplaced in fairly stony soils. Chalk soils are soft enough for penetration by a sharp augur to considerable depths.

Disadvantages include: (1) Some variation between individual blocks in a batch, and between batches occurs, probably during manufacture, and absolute individual calibration in tension units is difficult. (2) The readings are influenced by soil temperature variations and salt concentrations—both minor in our soils. (3) Some workers consider blocks rather insensitive to small changes of tension at the wet end. (4) Very rocky soils make deep emplacement difficult. (5) In some conditions, the electrical path between the electrodes can pass beyond the plaster envelope, and the block may then show a misleading response.

Examination of several years results from a range of soils and sites has given confidence in the use of blocks. It is considered that useful information can be obtained on the depths of abstraction of water, the times of the more intensive drying and wetting, and the progress and intensity of these processes. When used in conjunction with much less frequent gravimetric determinations, and Neutron count data, a more detailed picture of short-period changes can be obtained. See Hinson, Kitching & Fourt (1966).

W.H.Hinson, H.Gunston & D.F.Fourt

A.C. resistance bridges for field use. The measurement of the resistance of soil moisture units or the conductivity of soil pastes must be carried out with a.c. to minimize the effects of polarization. The bridges also enable one to use thermistors for temperature measurement.

Two designs of bridge were shown. The simple one (Hinson & Kitching, 1964) can be constructed by anyone who uses a soldering iron in a few hours at a cost of around £4. It uses headphones as the 'null' point indicator. The balance is reasonably sharp but depends on the state of the electrodes in the blocks.

The more elaborate bridge (Kitching, 1965) is particularly accurate and easy to use. It is an improvement on portable commercial equipment as it measures the true resistance of a reactive impedance without recourse to simultaneous balancing of resistive and reactive arms of the bridge. A phase sensitive detector is used which has the further advantage of indicating in which direction to move the cursor.

W.H.Hinson, H.Gunston & D.F.Fourt

The automatic recording of soil moisture—a method of coupling bouyoucos resistance blocks to a solartron data logger. A circuit is described which enables the soil moisture characteristics at 20 points to be sequentially recorded in digital form within 40 sec at pre-set time intervals. Two switched ranges of resistance are provided, the response curves of which are superimposed (Fig. 3).

The two bridge circuits (Fig. 5) are switched by Sl into the signal leads which connect the scanner unit to the command range unit, the

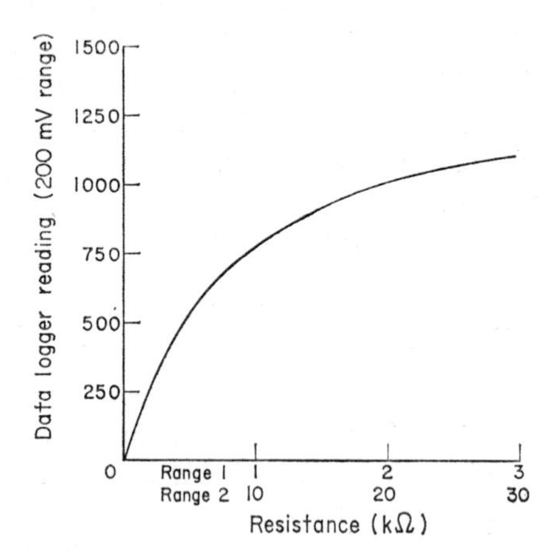

FIG. 3. Calibration curve.

resistance blocks being connected direct to the input plug of the scanner unit. This arrangement enables up to 20 resistance blocks to be used in a common bridge circuit.

A square wave of approximately 10 c.p.s. is generated by the multivibrator circuit Tr 1, Tr 2. This is amplified by Tr 3 and Tr 4, causing the relay coils L_1–L_4 to be alternately energized and de-energized. The magnetically operated reed switches, K_1–K_4, mounted inside the coils, are thus vibrated at the same frequency. K_1 and K_2 operate in such a manner that an alternating voltage appears across the bridge circuits. The peak value of this voltage depends on the setting of potentiometer RV_3, which provides a convenient means of adjustment of the full-scale reading. The lower end of the scale is set on the two ranges separately by adjustment for zero output of RV_1 and RV_2 with R_3 and R_9 short-circuited.

The alternating output from the selected bridge is applied to K_3

and K_4, which, being synchronized with K_1 and K_2, provides a d.c. voltage suitable for feeding into the digital voltmeter of the data logger. This method of rectification was used in preference to any arrangement using semiconductor rectifiers because it imposes no resistive load on the output of the bridge circuit.

A drying curve for one of the Bouyoucos (Bouyoucos, 1954) blocks exposed to a gentle air stream at 20°C is shown in Fig. 4.

Work is in progress to extend the range of the system.

F.Sutton & I.H.Rorison

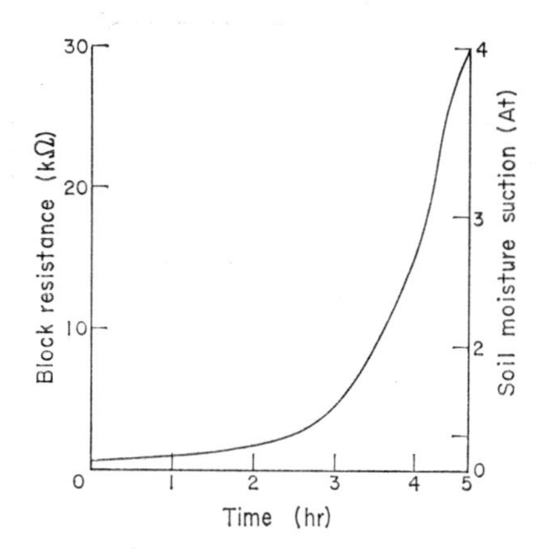

FIG. 4. Drying curve of Bouyoucos block exposed to a gentle air stream.

The study of root distribution in soil sections. The discovery of marked dis-equilibrium between aeration conditions in the gas and water phases at any one depth in wet-heath soils led to this investigation of the distribution of roots in relation to pore size. Its purpose, which was not precisely achieved, was to discover whether roots were in gas or water-filled pores. The method of preparation of soil sections was essentially that of Burges and Nicholas (1961). A blue pigment incorporated in the resin, enabled pores to be differentiated from transparent minerals without the use of a polarising microscope.

Undisturbed soil samples were collected in metal cylinders and dehydrated by freeze drying. They were then impregnated with Crystic 195 (Scott, Bader and Co. Ltd.) in which the blue pigment, also supplied by Scott Bader, was incorporated. During impregnation

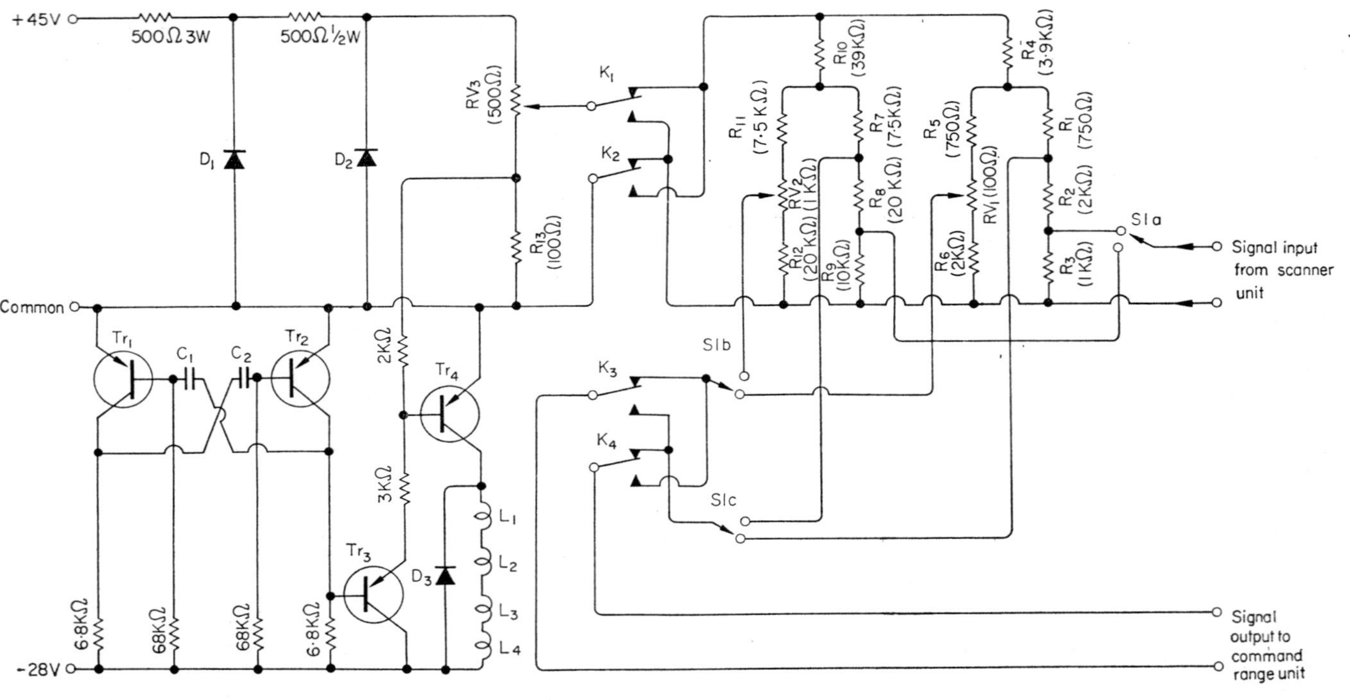

FIG. 5. Resistance measuring circuit. Resistors R_1 to R_{13} are high stability ± 1 per cent, other resistances are $\frac{1}{4}$ watt ± 5 per cent except where shown otherwise. Capacitors C_1 and C_2 are 2 μF, 63 V working, paper. Diodes D_1 and D_2 are types 1Z6.2 and 1Z18 of International Rectifier; D_3 is Mullard OA81. Transistors are Mullard ACY20. Switches K_{1-4} are Radiospares 13-RSR with (L_{1-4}) Radiospares coil type 1. Switch $S1_{a-c}$ is a 3 pole 2 way wafer switch.

the soil sample stood in a shallow dish and the pressure was progressively reduced during 30 min to about 15 cm mercury. This pressure was maintained for a further 30 min and was then gradually returned to atmospheric. The samples were left for about three weeks for the resin to mature and for the samples to become rock hard, before preparation of sections by the Imperial College Geology Department.

The distribution of roots in relation to pore diameters was studied on random transects across sections in the horizontal plane. No roots of *Molinia caerulea* or *Erica tetralix* were found in pores of less than 150 μ diameter and most were restricted to pores exceeding 300 μ diameter. If the pores were parallel sided those exceeding 150 μ diameter should be drained by tensions of 20 cm water. But a vertical section showed that the large pores were limited in vertical extent and opened through narrow necks. Some results obtained in soil were as follows:

(1) Pores exceeding 150 μ diameter in horizontal section: 50 per cent of total horizontal area.

(2) Water removed from saturated soil by 20 cm tension: 10 per cent of total soil volume.

(3) Continuous pores exceeding 150 μ diameter in vertical section: 6 per cent of total vertical area.

(4) Water removed from saturated soil by 80 cm tension: 18 per cent of total soil volume.

(2) and (3) are reasonably consistent and suggest that much of the pore space which contains roots, (1), remains water-filled when the water-table is 20 cm below the soil horizon. Even the water removed at 80 cm, (4), is only one-third of the root-inhabited pore space. In the soils studied, where the water table rarely fell more than 40 cm below the soil surface, it was concluded that a high proportion of the roots were in water-filled rather than gas-filled spaces. The analysis of sections for this purpose is difficult, and as yet incomplete and only semi-quantitative. **K.H.Sheikh & A.J.Rutter**

Soil boring and coring tools. *The Jarrett auger.* This Australian device should be better known as it provides a remarkably quick and easy method of cutting a hole—if necessary to a considerable depth. We use it for exploration and for making deep bores for many purposes. We have bored to 6 m in a light soil in under an hour and been through 2·5 m of gravel in the same time. We would like to hear from anyone wishing to purchase these tools as we can give them current information on the supply position.

The Hardypick Auger. We use this for similar purposes to the Jarrett but it is slower in most soils and does not 'eat' stones like the Jarrett. It leaves a particularly accurate hole which we prefer for use as a test well in studies of surface water on sites of experimental drainage. It is easier to control in making an inclined hole for emplacing instruments. The supplier is: Hardypick Ltd.

Modified scotch-eyed augers. We use these augers, which are manufactured for use in timber construction work, in a variety of sizes for emplacement of instruments, sometimes in the bottom of inclined holes cut with the larger tools. We are aware that these augers have been used for exploration in soil surveying but consider them inferior for this purpose to the larger tools; they provide only a smear of soil material of uncertain origin from which it is unwise to draw definite conclusions. We like to stand and watch the inexperienced use these augers in stony soils—especially if the results convince them that they should use other methods.

Large soil corers. These are of our own devising and represent an attempt to make manual tools to bring up volumetric samples or intact cores. We have achieved more or less continuous cores down to 6 ft with tools of this sort. The quality of the core varies with soil texture and moisture content. The head is of a special shock resisting chrome steel with a high content of tungsten and chromium. An occasional small flint will knap across as if cut in the true sense; quite large tree roots cut well and enter the core with little disturbance. However, this method of investigation is not suitable for stony soils and we have never fully solved the problem of retaining the core in the tube under exceptionally wet or dry conditions. Under some circumstances in the forest it may be the only possible way of getting a core without producing unacceptable disturance of the site. It works well on moist loams with a maximum bite of 45 cm. It can be very heavy work both going down and raising it. We would not wish to give the impression that we are recommending it for routine use, although it is good exercise.

W.H.Hinson, H.Gunston & D.F.Fourt

Soil corer. In principle this is a simple Kullenberg sampler, used to enable pH, redox potential and temperature readings to be taken down a wet soil profile. As the redox potential in such soils is often negative, and poorly poised, the soil must be removed without contact with the air. A Perspex tube, sharpened at one end, has two vertical rows of holes, sealed with strips of PVC insulating tape. A piston is positioned at the bottom of the tube, resting on the soil, and is held there while the tube is pushed into the soil. The vacuum under the piston draws the soil into the corer with very little compression, and holds the core in place when the corer is removed from the soil. At each depth in turn the holes are opened and an E.I.L. combined pH and reference electrode, a bright platinum wire electrode or a thermometer is inserted. A redox-pH switch is attached to the E.I.L. portable pH meter to facilitate rapid readings.

The corer should be of use in waterlogged or damp soft soils, providing there are few stones or woody roots and rhizomes. However, the escape of oxygen from reed rhizomes during sampling may change negative redox potentials. **D.F.Westlake**

Fifteen bar ceramic plate extractor. Pressure membrane extractor. Pressure plate extractor. Soil core sampler and soil retainer assembly. Soil water sampler. Soil tensiometers including gauge for recording at five depths simultaneously. Neutron scatter soil moisture apparatus. **Pitman**

ANALYSIS OF GAS IN THE ATMOSPHERE AND IN THE SOIL ATMOSPHERE

Oxygen monitor and oxygen meters by Yellow Springs Instrument Co. which measure oxygen content in liquids using probes with teflon membranes and oxygen in air using a cell separated from the air by a teflon membrane. **Shandon Scientific**

Portable CO_2 and temperature outfit. **Cambridge Instrument**

Probes for sampling gas and water in semi-waterlogged soils. The probes are constructed of Perspex cylinders, internal diameter 4·5 cm, and in lengths of 10, 15 and 30 cm. In use they are left *in situ*, and are sampled periodically.

The gas-sampling probes incorporate a chamber of approximately 60 ml volume. The gas initially in this chamber can equilibrate with the gas in the soil through a perforated plate. The idea is not new (see e.g. Conway, 1936), and is mentioned here for comparison with the water-sampling probe.

In the water-sampling probe the lower end is closed with a membrane of Porvic grade M (Porous Plastics Ltd.) supported on a perforated plate. Before insertion in the soil the pores of the Porvic are saturated with water under reduced pressure. After insertion a layer of medicinal paraffin and a drop of streptomycin solution is run into the probe. When a sample of soil water is required, pressure is reduced within the probe by about 10 cm mercury (measured with a manometer). Within a few hours sufficient water for analysis is withdrawn from the soil. Gas exchange between the water and the air in the probe is prevented by the floating layer of paraffin and the streptomycin prevents microbial respiration in the water. The sample is withdrawn through a long-stemmed pipette, is stored in a small bottle under paraffin, and is analysed within 48 hr. With this reduction in pressure water can be readily withdrawn from soil when the water table is within 40 cm of the sampling depth.

In the soils investigated, gas and water at a given depth were not in equilibrium with one another. The water phase, as might be expected, contained less oxygen and more carbon dioxide than would be the case if it were in equilibrium with the soil gas.

K.M.Sheikh & A.J.Rutter

A new chamber for the measurement of photosynthesis. A new chamber for gas exchange measurements on plants in the field as well as in the laboratory.

This chamber provides an accurate temperature as well as humidity control. It is possible to set the apparatus so that the environmental conditions inside the chamber correspond exactly with those outside. It is also possible to maintain a defined environmental condition inside independent from outside conditions. Finally it is possible to set an environmental control by a computed programme.

The chamber uses six peltier-elements for cooling and heating. The temperature is controlled with thermistors with an error of 0·1°c. Humidity is controlled with LiCl sensors which regulate a peltier cooled water vapour trap with an error of 0·5°c dewpoint in long term experiments (in short term experiments the error is 0·05°c). Wind speed is controlled with an infinitely variable fan.

E.-D.Schulze

Portable thermal conductivity gas analyser. Automatic, multipoint, gas sampling apparatus. This was designed to study the distribution of fumigants such as methyl bromide but is applicable to carbon dioxide and certain other gases. **H.E.Wainman**

RECORDING

Microscan data logger, IBM output writer and associated equipment. **Dynamco**

Compact logger. This equipment provides, at comparatively low cost, automatic recording in digital form of a number of d.c. voltage inputs from, for example, thermocouples, resistance thermometers etc. The printed output gives non ambiguous and accurate visual records. Punched paper tape gives facilities for subsequent re-play or computer analysis. **Solartron**

Incremental digital magnetic tape recorder (Meteorological Office type). **Lintronic**

Multipoint recorder fitted with a digitizer and having an external decoding unit. The decoder can be arranged to operate external tape punches and or typewriters. **Kent**

Model 25 tape punch for the conversion of parallel-wire (simultaneous) electrical impulses into punched paper tape. **Creed**

Miniature and multipoint potentiometric chart recorders. These are mains operated and transistorized. **Control Instruments**

DVX 315 0·005 per cent integrated digital voltmeter with plug-in units. 10 μV resolution (or 0·1 μV with pre-amplifier). **Advance**

Model 12-0147-00 floating decimal point printer. The decimal point can be inserted at any point in a group of digits. Printing can be in black or red. Other digital printers available. **Addo**

Strip chart potentiometric recorder. Recording oscillograph. X-Y plotter. **Honeywell**

Cheap photographic recorder for manometers and digital indicators, e.g. Seimens meter (Kipp thermopile) and cup contact anemometers.

Ex-government surplus 16 mm cine titling camera (cost £3 15*s.*) is triggered by solenoid-operated mercury switch. About six frames are exposed as the mercury flows past a pair of contacts. Coverage can be readily altered from about 5 cm × 5 cm to about 30 cm × 30 cm. **E.J.Winter**

See also pp. 255, 257–8, 264, 266, 268, 276–8, and 285–6.

MISCELLANEOUS

An air sampling unit for integrated environmental measurement. This device attempts to alleviate the problems of spatial variation in environmental sampling by replicating the sampling points. Air is drawn from four points through side-arms to a chamber where the flow is sufficiently turbulent to result in thorough mixing. This air then passes over wet and dry bulb resistance thermometers whose long axes lie parallel to the flow and from here is piped away for CO_2 analysis. Radiation shielding is necessary. **G.B.K.Baines**

A microplot technique to study the effect of soil-moisture conditions on the breakdown of woodland leaf litter. A major difficulty in the study of the effects of environmental factors on the fauna decomposing leaf litter on the forest floor is the variability of the litter environment under natural conditions. For this reason a microplot technique has been designed to permit studies under more uniform conditions. An effort has been made to devise a flexible structure to permit investigations of the effects of moisture, temperature and other factors on the litter fauna.

A series of concrete microplots one metre square have been constructed, each having a leaf-litter layer (*ca.* 5 cm deep) resting on a sand substrate maintained at one of four moisture levels obtained by adjusting the height of the water table in the sand and/or fine-gravel substrate. The object in the first instance is to determine the composition of the faunal population together with changes as the litter layer age and if possible to relate these to some of the physical properties of the environment. The experiment is designed to provide a balance sheet covering changes in the litter layers particularly those relating to moisture and temperature (measured by the rate of inversion of sucrose in an acid-buffer solution). **P.W.Murphy & C.J.D.Shackles**

Prototype combined photometer, thermometer and hygrometer for field use. Battery operated. **Flatters & Garnett**

A sampler to take core samples 40 cm long and 8 mm in diameter from sacks of flour. Damage to the sack is minimal and mixing along the length of the core is negligible. **G.A.Brett**

Grassmeter designed to make rapid non-destructive estimates of the yield of growing pasture. Swards at least 18 cm × 18 cm are needed but large plots can also be used (Alcock and Lovett, 1967). **M.B.Alcock**

Clip-on diffusion porometer. This porometer is of the design suggested by Van Bavel, Nakayama and Ehrler (1965) but has certain design improvements. The speed of change of humidity in a chamber held against the leaf is measured. A brief description of this instrument will be found in Meidner and Mansfield (1968). **W.Stiles**

A method of producing soil crusts and measuring their modulus of rupture. A method is described for making experimental soil crusts, of different 'strengths', and subjecting samples of the crusts to destructive tests to measure the modulus of rupture of areas of different sizes (Collip and Heydecker 1967). The technique is of particular value in connection with work on the factors that affect the emergence of germinating seedlings under field conditions.

H.F.Collip & W.Heydecker

Hydrometeorological station. Developed for catchment area research in remote regions where mains power supplies are not available, the Hydrological Research Unit automatic hydrometerorological station provides, at modest cost, a means of recording climatic data to 1 per cent in a form suitable for automatic handling from start to finish. The sampling interval is six minutes: the resulting daily mean of 240 instantaneous values is very close to the true daily value.

The data logger used is the 'Limpet Logger' (d-mac Ltd.) which records 10 analogue inputs as a series of pulses on 6 cm ($\frac{1}{4}$in.) magnetic tape, it operates from dry batteries for an extended period and can take 35,000 recordings per side of the tape. Resolution is limited to 100 discrete levels and a compromise between range and resolution is therefore necessary. For any single parameter, such as rainfall where an increase in the number of steps is necessary an additional channel may be used but in general, discrimination is adequate for most hydrological purposes.

Sensors and transducers were described recently by Strangeways & McCulloch (1965) and measure solar radiation, net radiation, temperature, humidity, rainfall, windrun, wind direction and time. Other parameters such as stream level, evaporation pan level, water temperature in evaporation pan, soil temperature, stream turbidity, dissolved oxygen and electrical conductivity, may be incorporated in the system. However, in the first instance, standard stations only are available.

The manufacturers of the logging equipment provide a console main translator which gives either typewriter or punched tape output. However, if any sizeable number of such stations is contemplated, a small computer is desirable. A service of translation and computation for standard stations will be offered by the Hydrological Research Unit on payment.

The Hydrological Research Unit hydrometeorological station has been in operation since mid 1965 and Marks I, II, and III have been produced and operated prior to the device being production engineered for commercial exploitation early in 1967. The first batch of ten production stations will be used in pairs for six months in a comprehensive test programme at sites where reliable conventional measurements of the appropriate parameters are available and will include tropical testing. Subject to satisfactory completion of these tests the equipment will be available after October 1967 from: Rock and Taylor Ltd. **I.C.Strangeways & J.S.G.McCulloch**

The measurement of volatile ammonia loss from soil. This simple apparatus is identical in principle to others used to measure ammonia loss from soil (Gasser, 1964) and is constructed with conventional laboratory glass and rubber components.

A sample of moist soil (100 g; exposed surface area *c* 50 cm^2; 2·5–3·0 cm depth) is contained in a 250 ml conical flask. Air moistened by bubbling through distilled water held in a 120 ml bottle is drawn by suction at the rate of *c* 2 l/min through the experimental flask and volatile ammonia is swept into 25·0 ml of 0·1 N H_2SO_4 contained in a second 120 ml bottle. Every third day the acid in the trap is titrated against 0·1 N NaOH using screened methyl orange indicator (Vogel, 1961). Contaminating ammonia in the laboratory atmosphere can be detected by substituting dilute acid for water in the bottle used to moisten the inflowing air.

Techniques to measure ammonia loss from soil under field conditions have been described (see Table 1) and they are all closed systems e.g. bell-jars, in which acid is used to trap evolved ammonia.

Volatile ammonia (gaseous NH_3) can be formed in soil during the decomposition of nitrogen-rich organic residues (Smith, 1964) and during the hydrolysis of urea voided in urine or applied as a fertilizer, e.g. urea or ammonium sulphate. Field measurements of volatile loss

TABLE 1. Volatile loss of ammonia from fertilizer urea applied to soil

Location	Urea rate lb N/acre	Ammonia loss lb N/acre	Authors
USA—Florida	100	59	Volk, 1959
USA—Pennsylvania	300	70	Kresge & Satchell, 1960
Malaya	200	50	Watson *et al*, 1962
Ghana	1250	375	Acquaye & Cunningham, 1965

of ammonia nitrogen from urea applied to soil surfaces illustrating maximum losses are shown in Table 1.

Doak (1952) found that sheep urine patches on pasture contain the equivalent of *c* 430 lb N/acre but no estimate seems to have been made of ammonia loss from urine patches in the field. The conditions which appear to favour ammonia loss from the surface layers of soil are high soil pH, soil drying, high soil temperatures and a low ammonium absorption capacity. Depending on factors such as soil structure and moisture content little volatile ammonia is lost when it is formed (or placed as a fertilizer) within the soil.

Besides contributing to loss of nitrogen from soil systems volatile ammonia can be toxic to plants and to nitrifying bacteria (Court, Stephen & Waid, 1964). The effects of volatile ammonia on organisms living in soil has been largely ignored as a factor in biological interactions amongst soil organisms even though it is inhibiting to many micro-organisms and can act as an attractant for certain invertebrates.

K.B.Pugh & J.S.Waid

A.C. resistance bridges for field use. See pages 255, 256 and 275.

Integrators for small electric currents. See pages 256, 258 and 259.

REFERENCES

Acock B. (1967) Thermistor leaf clip and sensitive multi-range thermistor thermometer. *University of Nottingham Dept. of Horticulture Misc. Pub.* 21.

Acquaye D.K. & Cunningham R.K. (1965) Losses of nitrogen by ammonia volatilization from surface-fertilized tropical forest soils. *Trop. Agric. Trin.* **42**, 281–92.

Alcock M.B. & Lovett J.V. (1967) The electronic measurement of the yield of growing pasture. *J. Agric. Sci., Camb.* **68**, 27–38.

Anderson M.C. (1964) Studies in the woodland light climate. I. The photographic interpretation of light conditions. *J. Ecol.* **52**, 27–41.

Blackwell M.J. (1953) On the development of an improved Robitzsh-type actinometer. *Met. Res. Ctee., Air Ministry, Lond., Met. Res. Pamphl.* **791**, 1–10.

Borhan A. & Acock B. (1967) Portable recording psychrometer mast. *Univ. of Nottingham Dept. of Horticulture Misc. Pub.* 19.

Bouyoucos G.J. (1954) New type electrode for plaster of paris moisture blocks. *Soil Sci.* **78**, 339–42.

Burges A. & Nicholas D.P. (1961) Use of soil sections in studying amount of fungal hyphae in soil. *Soil Sci.* **92**, 25–9.

Collip H.F. & Heydecker W. (1967) Method of producing soil crusts and measuring their strength. *Univ. of Nottingham Dept. of Horticulture, Misc. Pub.* 23.

Conway V.M. (1936) Studies in the autecology of *Cladium mariscus* R.Br. II. Environmental conditions at Wicken Fen, with special reference to soil temperatures and the soil atmosphere. *New Phyt.* **35**, 359–80.

Coombe D.E. & Evans G.C. (1966) Hemispherical photography in studies of plants. In: *Light as an Ecological Factor.* (Ed. R. Bainbridge, G.C. Evans and O. Rackham), p.417. Blackwell, Oxford.

Court M.N., Stephen R.C. & Waid J.S. (1964) Toxicity as a cause of the inefficiency of urea as a fertilizer. I. Review. *J. Soil Sci.* **15**, 42–8.

Doak B.W. (1952) Some chemical changes in the nitrogenous constituents of urine when voided in pasture. *J. agric. Sci., Camb.* **42**, 162–71.

Evans G.C. (1966a) Model and measurement in the study of woodland light climates. In: *Light as an Ecological Factor.* (Ed. R.Bainbridge, G.C.Evans and O.Rackham), pp.53–76. Blackwell, Oxford.

Evans G.C. (1966b) Apparatus for surveying light climate in woodlands, including area and intensity of sunflecks. In: *Light as an Ecological Factor.* (Ed. R.Bainbridge, G.C.Evans and O.Rackham), pp.418–20. Blackwell, Oxford.

French C.S. (1960) The chlorophylls *in vivo* and *in vitro*. In: *Hanbuch der Pflanzenphysiologie*, V1, 252–97. Ed. W.Ruhland, Springerverlag, Berlin.

Gasser J.K.R. (1964) Some factors affecting losses of ammonia from urea and ammonium sulphate applied to soils. *J. Soil Sci.* **15**, 258–71.

Hand D.W. (1967) An electrically-weighed lysimeter based on the use of strain gauges. *Univ. Nottingham Dept. of Horticulture, Misc. Pub.* 20.

Head M.R. & Surrey N.B. (1965) Low speed anemometer. *J. scient. Instrum.* **42**, 349.

Head M.R. & Thorpe R.R. (1965) Direct reading low speed anemometer. *J. scient. Instrum.* **42**, 811.

Hinson W.H. & Kitching R. (1964) A readily constructed transistorized instrument for electrical resistance measurements in biological research. *J. app. Ecol.* **1**, 301–5.

Hinson W.H., Kitching R. & Fourt D.F. (1966) Soil moisture, climate and tree growth. *Rep. Forest Res., Lond.* p.60–2.

Hirst J.M. (1957) A simplified surface witness recorder. *Pl. Path.* **6**, 57–61.

Hirst J.M., Long I.F. & Penman H.L. (1954) Micrometerology in the potato crop. *Proc. met. Conference, Toronto* 1953, 233–7.

Hudson J.P. (1967) Drainage lysimeters operated by remote control. *Univ. of Nottingham Dept. of Horticulture, Misc. Pub.* 22.

Kitching R. (1965) A precision portable electrical resistance bridge incorporating a centre zero null detector. *J. agric. Engng. Res.* **10**, 264–6.

Kresge C.B. & Satchell D.P. (1960) Gaseous loss of ammonia from nitrogen fertilizers applied to soils. *Agron. J.* **52**, 104–7.

Macfadyen A. (1956). The use of a temperature integrator in the study of soil temperature. *Oikos*, **7**, 56–71.

Meidner H. & Mansfield T.A. (1968) *The physiology of stomata.* McGraw Hill, New York.

Monteith J.L. (1959) Solarimeter for field use. *J. scient. Instrum.* **36**, 341–6.

Monteith J.L. & Szeicz G. (1962) Simple devices for radiation measurement and integration. *Arch. Met.* **11**, 491–500.

Post J.J., Allison C. C., Burckhardt H.& Preece T. F. (1963) The influenceof weather conditions on the occurrence of apple scab. *Tech. Notes Wld. met. Org.* **55**, pp.41.

Powell M.C. & Heath O.V.S. (1964) A simple and inexpensive integrating photometer. *J. exp. Bot.* **15**, 187–91.

Preece T.F. (1964) Continuous testing for apple scab infection weather using apple rootstocks. *Pl. Path.* **13**, 6–9.

Preece T.F. & Smith L.P. (1961) Apple scab infection weather in England and Wales, 1956–60 *Pl. Path.* **10**, 43–51.

Smith J.H. (1964) Relationships between soil cation-exchange capacity and the toxicity of ammonia to the nitrification process. *Soil Sci. Soc. Amer. Proc.* **28**, 640–4.

Somerville M.T. & Turnbull G.F. (1963) Self-generating high frequency carrier-feedback anemometer. *Proc. Instn. elect. Engrs.* **110**, 10.

Strangeways I.C. & McCulloch J.S.G. (1965) A low cost automatic hydrometeorological station. *Bull. Int. Assoc. Scient. Hydrol.* Xe anne, 57–62.

Szeicz G. (1965) A miniature tube solarimeter. *J. appl. Ecol.* **2**, 145–7.

Szeicz G., Monteith J.L. & Dos Santos J.M. (1964) Tube solarimeter to measure radiation among plants. *J. appl. Ecol.* **1**, 169–74.

Taylor G.I. (1927) Bi-directional wind-vane. *Q. Jl. R. met. Soc.* **53**, 201.

Thompson B.W. (1953) Aircraft applications of insecticides in East Africa. III. Atmospheric turbulence in woodland. *Bull. ent. Res.* **44**, 611–26.

Vogel A.I. (1961) *Quantitative Inorganic Analysis.* 3rd Ed. Longmans, Lond.

Volk G.M. (1959) Volatile loss of ammonia following surface application of urea to turf or bare soils. *Agron. J.* **51**, 746–9.

Van Bavel C.H.M., Nakayama F.S. & Ehrler W.L. (1965) Measuring transpiration resistance of leaves. *Plant Physiol., Lancaster*, **40**, 535–40.

Warren Wilson J. (1965) Stand structure and light penetration. I. Analysis by point quadrats. *J. appl. Ecol.* **2**, 383–90.

Watson G.A., Chin Tet Tsoy & Wong Phui Weng (1962) Loss of ammonia by volatilization from surface dressings of urea in *Hevea* cultivation. *J. Rubb. Res. Inst. Malaya.* **17**, 77–90.

Weaving G.S. (1967) A portable thermistor temperature integrator. *J. scient. Instrum.* **44**, 55.

ADDRESSES OF FIRMS

The firms listed here are those who gave a demonstration or who were mentioned by demonstrators

Addo Ltd, Automatic Data Processing Division, 8 Greyfriars Road, Reading, Berkshire
Advance Electronics Ltd, Roebuck Road, Hainault, Ilford, Essex
Baird and Tatlock (London) Ltd, Freshwater Road, Chadwell Heath, Essex
Cambridge Instrument Co Ltd, 13 Grosvenor Place, London SW1
C.F.Casella and Co Ltd, Regent House, Britannia Walk, London N1
CIBA (A.R.L.) Ltd, Duxford, Cambridge
Control Instruments Ltd, Alfreton Road, Derby
Creed and Co Ltd, Hollingbury, Brighton, Sussex
F.Darton and Co Ltd, Mercury House, Vale Road, Bushey, Watford, Herts
d-mac Ltd, 55 Kelvin Avenue, Glasgow SW2
Drums Ltd, Grosvenor Gardens House, London SW1
Dynamco Systems Ltd, Govett Avenue, Shepperton, Middlesex
Electronic Instruments Ltd, Richmond, Surrey
Flatters and Garnett Ltd, Mikrops House, Bradnor Road, Manchester 22
Foster Instrument Co Ltd, Letchworth, Herts
Grant Instruments (Developments) Ltd, Toft, Cambridge
Hardypick Ltd, Archer Road, Millhouses, Sheffield 8
Honeywell Controls Ltd, Great West Road, Brentford, Middlesex
Kelseal Ltd, Vogue House, Hanover Square, London W1
George Kent Ltd, 6 Hunting Gate, Hitchin, Herts
Lintronic Ltd, 54–58 Bartholomew Close, London EC1
Claud Lyons Ltd, Instruments Division, Ware Road, Hoddesdon, Herts
Megatron Ltd, 115A Fonthill Road, London N4
D.A.Pitman Ltd, 91 Heath Road, Weybridge, Surrey
Plessey Automation, Systems Development Division, Sopers Lane, Poole, Dorset
Porous Plastics Ltd, Dagenham Dock, Essex
Radiospares Ltd, P.O. Box 2BH, 4–8 Maple Street, London W1
Rock and Taylor Ltd, Hayes Lane Trading Estate, Lye, Stourbridge, Worcestershire
Scott, Bader and Co Ltd, Wollaston, Wellingborough, Northants
Shandon Scientific Co Ltd, 65 Pound Lane, Willesden, London NW10
Short and Mason Ltd, Wood Street, London E17
The Solartron Electronic Group Ltd, Farnborough, Hampshire
Standard Telephone and Cables Ltd, Thermistor Division, Footscray, Sidcup, Kent

LIST OF THOSE WHO ATTENDED THE SYMPOSIUM

Acock Mr B. *Horticulture Department, University of Nottingham, Sutton Bonington, Loughborough.*
Adams Dr W.A. *Soil Science Unit, University College of Wales, Aberystwyth*
Akande Mrs M. *Department of Forestry and Natural Resources, University of Edinburgh, 10 George Sq., Edinburgh 8.*
Alcock Mr M.B. *Department of Agriculture, U.C.N.W., Bangor, N. Wales.*
Allaway Mr W.G. *Biology Department, Lancaster University, St. Leonard's House, Lancaster.*
Amos Mr K.J. *Standard Telecommunication Laboratories, London Rd., Harlow, Essex.*
Amos Dr T.G. *Department of Natural History, Queen's College, Dundee.*
Andersson Mr F. *Department of Plant Ecology, Ö. Vallgatan 14–18, Lund, Sweden.*
Andrews Miss R.E. *Botany Department, University College of Wales, Aberystwyth.*
Anslow Mr R.C. *Grassland Research Institute, Hurley, nr. Maidenhead, Berks.*
Aram Mr R.H. *School of Biological Sciences, University of Sussex, Falmer, Brighton.*
Arnold Miss M.K. *284 The Ridgeway, St. Albans, Herts.*
Arnott Mr R.A. *Grassland Research Institute, Hurley, nr. Maidenhead, Berks.*
Ashby Dr M. *58 Gains Road, Southsea, Hants.*
Ashman Mr F. *'Hollyside', Parish Lane, Farnham Common, Slough, Bucks.*
Aston Dr J.L. *Unilever Research Laboratory, Colworth House, Sharnbrook, Beds.*
Atkinson Mr B.W. *Department of Geography, Queen Mary College, Mile End Road, London, E.1.*
Avery Dr D.J. *East Malling Research Station, Nr. Maidstone, Kent.*
Bacon Mr D.W.G. *2 The Ramparts, Sandwich, Kent.*
Badcock Miss R.M. *Department of Biology, The University, Keele, Staffs.*
Bailey Mr S.W. *c/o Pest Infestation Laboratory, London Road, Slough, Bucks.*
Baines Mr G.B.K. *School of Agriculture, University of Nottingham, Sutton Bonington, Loughborough.*
Baker Mr C.R.B. *M.A.F.F. Plant Pathology Lab. Hatching Green, Harpenden, Herts.*
Banage Dr W.B. *Imperial College Field Station, Sunninghill, Ascot, Berks.*
Bannister Dr. P. *Department of Botany, The University, Glasgow, W.2.*
Batten Miss A. *Anti-Locust Research Centre, Wrights Lane, London, W.8.*
Beaumont Mrs E.M. *190 Balfour Road, Brighton, Sussex.*
Bell Mr J.N.B. *Department of Botany, University of Manchester, Manchester, 13.*
Bellamy Mr D.J. *Department of Botany, South Road, Durham City.*
Berrie Dr A.D. *Department of Zoology, University of Reading, Reading, Berks.*
Berry Mr W.G. *Department of Geography, 17 Bell Street, Dundee.*
Berthet Dr P. *Institut de Zoologie, 59 rue de Namur, Louvain, Belgium.*
Blackburn Mr M.R. *Meteorological Research Unit, Huntingdon Road, Cambridge.*
Blackwell Mr M.J. *Met. Office, Met. O. 16a, London Road, Bracknell, Berks.*
Bligh Dr J. *ARC Institute of Animal Physiology, Babraham, Cambridge.*
Block Dr W. *School of Agriculture, University of Cambridge, Downing St., Cambridge.*

Blofield Miss B.A. *5 Winchelsea Close, Chartfield Avenue, Putney, London, S.W.15.*

Bocock Mr K.L. *The Nature Conservancy, Merlewood Research Station, Grange-over-Sands, Lancs.*

Boga Miss D.S. *Anti-Locust Research Centre, College House, Wrights Lane, London, W.8.*

Borhan Dr A. *School of Agriculture, Sutton Bonington, Loughborough.*

Bowman Mr G.E. *N.I.A.E., Wrest Park, Silsoe, Bedford.*

Brandt Dr D.Ch. *Zoologisch Laboratorium, Kaiserstraat 63, Leiden, Holland.*

Bremner Mr D. *Malham Tarn Field Centre, Nr. Settle, Yorks.*

Brett Mr G.A., *MAFF Infestation Control Laboratory, Gort Buildings, Hook Rise South, Tolworth, Surbiton, Surrey.*

Brian Mr M.V. *Furzebrook Research Station, Wareham, Dorset.*

Bridal Mr T.A. *'Haulfre', Park Crescent, Llanfairfechan, Caerns.*

Brindley Mr P. *Department of Agricultural Botany, University College of North Wales, Bangor, Caerns.*

Broadhead Dr E. *Zoology Department, The University, Leeds.*

Brockway Dr J.M. *Rowett Research Institute, Bucksburn, Aberdeen.*

Brook Mr D.W.I. *Department of Agriculture, The University, Reading, Berks.*

Brookes Mr D. *Hatfield College of Technology, Hatfield, Herts.*

Brookhouse Dr J.K. *University Statistician, The University, Reading, Berks.*

Brooks Dr D.H. *Jealott's Hill Research Station, Bracknell, Berks.*

Brown Mr J.M.B. *The Clock House, Dockenfield, Farnham, Surrey.*

Bullen Mr F.T. *Anti-Locust Research Centre, Wrights Lane, London, W.8.*

Bunt Mr A.C. *Glasshouse Crops Research Institute, Worthing Road, Littlehampton, Sussex.*

Bunting Prof. A.H. *Department of Agricultural Botany, The University, Reading, Berks.*

Burdekin Dr. D.A. *Alice Holt Lodge, Forestry Commission Research Station, Wrecclesham, Farnham, Surrey.*

Burnett Dr G.F. *Marischal College, University of Aberdeen.*

Burrage Dr S.W. *Wye College, Wye, Nr. Ashford, Kent.*

Burrell Mr N.J. *Pest Infestation Laboratory, London Road, Slough, Bucks.*

Burrows Dr F.J. *c/o Welsh Plant Breeding Station, Plas Gogerddan, Aberystwyth,*

Caborn Dr J.M. *Department of Forestry and Natural Resources, University of Edinburgh, 10 George Square, Edinburgh 8.*

Carrick Mr T.R. *Pest Infestation Laboratory, London Road, Slough.*

Carter Mr C.I. *Forestry Commission Research Station, Alice Holt Lodge, Farnham, Surrey.*

Carter Miss J. *Department of Geography, University of Manchester, Manchester 13.*

Caswell Mr B.B. *27 Danecourt Gardens, East Croydon, Surrey.*

Chadwick Dr M.J. *Department of Biology, University of York, Heslington, York.*

Chapas Mr L.C. *Grassland Research Institute, Hurley, Nr. Maidenhead, Berks.*

Chapman Dr R.F. *Zoology Department, Birkbeck College, Malet Street, London, W.C.1.*

Chapman Dr S.B. *The Nature Conservancy, Furzebrook Research Station, Wareham, Dorset.*

Chappell Mr H.G. *Department of Biological Sciences, Portsmouth College of Technology, Hay Street, Portsmouth.*

Chartier Mr P. 1 *rue Tristan Bernard, 78 Les Clayes, Bois, France.*
Chatfield Miss J.E. *Department of Zoology, Reading University, Reading, Berks.*
Chawner Mr P.M.H. *Hampden Test Equipment Ltd., 20–24 St. Andrews Street, Northampton.*
Chudley Mr J.L. *c/o Department of Education Science Laboratory, Reading University, Reading, Berks.*
Clymo Dr R.S. *Westfield College, Kidderpore Avenue, London, N.W.3.*
Colebourn Mr P.H. *Department of Biology, University of Lancaster, St. Leonardgate, Lancaster.*
Collingbourne Mr R. *Met. Office (Met O. 14), London Road, Bracknell, Berks.*
Collip Dr H.F. *School of Agriculture, Sutton Bonington, Loughborough.*
Coombe Dr D.E. *The Botany School, Downing Street, Cambridge.*
Coombs Mr C.W. *Pest Infestation Laboratory, London Road, Slough.*
Cooper Dr A.J. *Glasshouse Crops Research Institute, Worthing Road, Littlehampton, Sussex.*
Cooper Dr J.P. *Welsh Plant Breeding Station, Plas Gogerddan, Aberystwyth.*
Coupland Prof. R.T. *Department of Plant Ecology, University of Saskatchewan, Saskatoon, Sask., Canada.*
Cousens Mr J.E. *Department of Forestry and Natural Resources,* 10 *George Square, Edinburgh* 8.
Coveney Mr R.D. *Tropical Stored Products Centre, London Road, Slough, Bucks.*
Cranstoun Mr G.N. 11 *Auchengreoch Road, Johnstone, Renfrewshire.*
Critchley Mr B.R. *Imperial College Field Station, Silwood Park, Ascot, Berks.*
Crofts Mr R. *Department of Geography, St. Mary's, Old Aberdeen.*
Crosbie Dr A.J. *Department of Geography, High School Yards, Edinburgh.*
Crothers Mr J.H. *Dale Fort Field Centre, Haverfordwest, Pembs.*
Cussans Mr G.W. *The Weed Research Organization, Begbroke Hill, Kidlington, Oxford.*
Dale Dr M. *Botany Department, University of Hull.*
Daniels Mr R.E. *Department of Botany, University of Nottingham.*
Davies Mr A.J. *Agriculture (Crop Husbandry) Department, Institute of Rural Science, Penglais, Aberystwyth.*
Dean Mr C.G. *University of Nottingham, School of Agriculture, Sutton Bonington, Loughborough.*
Denny Mr P. *Botany Department, The University of St. Andrews.*
Dibley Mr G.C. *Rothamsted Experimental Station, Harpenden, Herts.*
Dickinson Miss W. 24 *The Fairway, Alwoodley, Leeds* 17.
Dixon Dr A.F.G. *Zoology Department, The University, Glasgow, W.2.*
Doley Mr D. *c/o Commonwealth Forestry Institute, South Parks Road, Oxford.*
Duffey Dr E.A.G. *Monks Wood Experimental Station, Abbots Ripton, Hunts.*
Eckardt Dr F.E. *CNRS, BP.* 1018, 34-*Montpellier, France.*
Eden, Mr M.J. *Bedford College, London, N.W.1.*
Edgar Mr W.D. *Glasgow University Field Station, Rowardennan, by Glasgow.*
Edington Dr J.M. *Department of Zoology, University College, Cathays Park, Cardiff.*
Edwards Dr R.S. *Department of Agriculture, U.C.W., Penglais, Aberystwyth.*
Egglishaw Dr H.J. *Freshwater Fisheries Laboratory, Pitlochry, Perthshire.*
Egziabher Mr T.B.G. *Department of Botany, University College of N. Wales, Bangor.*

Ellis Dr P.E. *Anti-Locust Research Centre, College House, Wright's Lane, London, W.8.*
Ellis Mr P.J. *Physics Department, The University, Reading, Berks.*
Elston Dr J.F. *Department of Agricultural Botany, The University, Reading, Berks.*
Etherington Dr J.R. *Botany Department, University College, Cardiff.*
Evans Dr G.C. *The Botany School, Downing Street, Cambridge.*
Fager Dr E.W. *Scripps Institution of Oceanography, Lajolla, California.*
Fairhurst Dr C.P., *Department of Zoology, Manchester University, Manchester* 13.
Farazdaghi Mr H. *Agriculture Department, The University, Reading, Berks.*
Flegg Mr P.B. *Glasshouse Crops Research Institute, Littlehampton, Sussex.*
Foster Dr J. 95 *Turf Hill Road, Rochdale, Lancs.*
Fourt Mr D.F. *Forest Research Station, Alice Holt Lodge, Farnham, Surrey.*
Fraser Mr A.I. *Forest Research Station, Alice Holt Lodge, Farnham, Surrey.*
Freeman Dr J.A. *M.I.F.F. Infestation Control Laboratory, Hook Rise, Tolworth, Surbiton, Surrey.*
French Mr R.A. *Rothamsted Experimental Station, Harpenden, Herts.*
Gardner Mr G. *Department of Botany, The University, Sheffield,* 10.
Gibbon Mr D.P. *Department of Agriculture, Leeds University.*
Gifford Dr D.R. *Department of Forestry and Natural Resources, University of Edinburgh,* 10 *George Square, Edinburgh* 8.
Giles Mr P.H. *Tropical Stored Products Centre, London Road, Slough, Bucks.*
Gloyne Dr R.W. *Meteorological Office,* 26 *Palmerston Place, Edinburgh* 12.
Goddard Mr I.C. *Flatters & Garnett Ltd., Wynnstay Grove, Manchester* 14.
Goode Mr D.A. 2 *Westbourne Avenue, Hull, Yorkshire.*
Goodman Mr G.T. *Department of Botany, University College, Swansea.*
Goodway Dr K.M. *Department of Biology, University of Keele, Keele, Staffs.*
Grainger Prof. J.N.R. *Department of Zoology, Trinity College, Dublin,* 2.
Grant Miss S.A. *Hill Farming Research Organisation,* 29 *Lauder Road, Edinburgh* 9.
Grebet Mr Ph. 59 *Avenue de la Bourolormois* 75, *Paris* 7e, *France.*
Green D.A. *Unilever Research Laboratory, Colworth House, Sharnbrook, Beds.*
Green Mr F.H.W. *Natural Environment Research Council, State House, High Holborn, London, W.C.*1.
Greig Mr B.J.W. *Forest Research Station, Alice Holt Lodge, Farnham, Surrey.*
Groenewoud Dr H. van, *Forest Research Laboratory Department of Forestry, Box* 4000, *Fredericton, N.B., Canada.*
Guild Dr W.J. *Department of Agricultural Zoology, University of Edinburgh, West Mains Road, Edinburgh* 9.
Gulliver Mr R.L. 40 *Crosbie Road, Harbourne, Birmingham,* 17.
Gunston Mr H. *Forestry Commission, Alice Holt Lodge, Farnham, Surrey.*
Hall Mr C.G. *Grant Instruments (Developments) Ltd., Toft, Cambridge.*
Halliday Miss T.A. *Department of Biological Sciences, Portsmouth Technical College, Hay Street, Portsmouth.*
Halstead D.G.H. *Pest Infestation Laboratory, London Road, Slough, Bucks.*
Hamilton Mr P.A. *Zoology Field Station, Rowardennan, by Glasgow.*
Hand Dr D.W. *N.I.A.E. Wrest Park, Silsoe, Beds.*
Harding Dr D.J.L. *East Malling Research Station, Maidstone, Kent.*
Harley Prof. J.L. *Department of Botany, The University, Sheffield,* 10.
Harris Mr P.M. *Department of Agriculture, The University, Reading, Berks.*
Hartland-Rowe Dr R. *Greenaway Lodge, Ashburton, Nr. Newton Abbot, Devon.*

Hartley Dr G.S. *Chesterford Park Research Station, Nr. Saffron Walden, Essex.*
Haslam Dr S.M. *Biology Department, Royal University of Malta, Malta.*
Hatto Miss J. *Department of Agricultural Botany, U.C.N.W., Bangor, N. Wales.*
Haynes Mr F.N. *Department of Biological Sciences, College of Technology, Hay Street, Portsmouth, Hants.*
Heal Dr O.W. *Merlewood Research Station, Grange-over-Sands, Lancs.*
Healey Dr I.N. *Department of Zoology, King's College, Strand, London, W.C.2.*
Heath Dr G.W. *Jacks Dell, Delmerend Lane, Hamstead, St. Albans, Herts.*
Heathcote Mr D.G. *Botany Department, University College, Cathays Park, Cardiff.*
Hemming Mr C.F. *Tudor Cottage, Long Crendon, Aylesbury, Bucks.*
Herliny Mr P.J. *Rhyd-y-Creua, The Drapers' Field Centre, Betws-y-Coed, Caerns.*
Hewett Mr D.G. 18 *Miles Avenue, Sandford, Wareham, Dorset.*
Heydecker Dr W. *School of Agriculture, Sutton Bonington, Loughborough.*
Heywood Mr R.B. 62 *Masefield Avenue, Borehamwood, Herts.*
Hinson Dr W.H. *Forest Research Station, Alice Halt Lodge, Farnham, Surrey.*
Hodek Dr I. *Entomological Institute, Praha* 2, *Viničná* 7, *Czechoslovakia.*
Hodge Mr C.A.H. 8 *Lilac Close, Haslingfield, Cambs.*
Hodgson Mr J. 1 *Louth Road, Sheffield* 11.
Hollobone Mr T.A. *N.R.D.C.,* 66 *Victoria Street, London, S.W.*1.
Holroyd Mr J. *A.R.C. Weed Research Organization, Kidlington, Oxford.*
Howard Mrs E.M. *River House, Piddinghoe, Newhaven, Sussex.*
Howe, Dr R.W. *Pest Infestation Laboratory, London Road, Slough.*
Howland Mr B.G. *c/o Glenbervie House, The Holt, Farnham, Surrey.*
Hubbard Mr J.C.E. *The Nature Conservancy, Furzebrook, Wareham, Dorset.*
Hudson Prof. J.P. *School of Agriculture, Sutton Bonington, Loughborough.*
Hughes Dr A.P. *A.R.C. Unit, Shinfield Grange, Shinfield, Berks.*
Hughes Dr J.C. *A.R.C. Food Research Institute, Earlham Laboratory, Recreation Road, Norwich, NOR* 26*G.*
Hughes Mr M.K. *Department of Zoology, South Road, Durham City.*
Hunham Miss N.M. 5 *Wentworth Road, Hertford, Herts.*
Hurlock Mr E.T. *M.A.F.F. Infestation Control Laboratory, Hook Rise South, Tolworth, Surbiton, Surrey.*
Hutnik Prof. R.J. 312 *Forestry Research Laboratory, The Pennsylvania State University, University Park, Pa., U.S.A.* 16802.
Huxley Mr T. *The Nature Conservancy,* 12 *Hope Terrace, Edinburgh* 9.
Hyde Miss M.B. *Pest Infestation Laboratory, London Road, Slough, Bucks.*
Ibbotson Dr A. *School of Agriculture, The University, Newcastle upon Tyne,* 2.
Idle Dr D.B. *Department of Botany, The University, Birmingham,* 13.
Ingram Mr P.R.P. *School of Biological Sciences, University of East Anglia, Wilberforce Road, Norwich, Nor* 77*H.*
Ingram Drs H.A.P. & R. *Department of Botany, Queen's College, Dundee.*
Is-Hag Mr H.M. *Department of Agricultural Botany, The University, Reading, Berks.*
Jackson Mr J. *Glasgow University Field Station, Rowardennan, Drymen, Glasgow.*
Jackson Dr J.E. *East Malling Research Station, Maidstone, Kent.*
Jacobs Mr L. *Meteorological Office (Met O.* 14), *London Road, Bracknell, Berks.*
Jager Mr J. de. *Welsh Plant Breeding Station, University College of Wales, Aberystwyth.*
James Mr D.B. *Department of Agricultural Botany, Institute of Rural Science, Penglais, Aberystwyth.*

James Miss K. 68 *Watford Road, Birmingham* 30.
James Mr R. *Department of Zoology, University College, Cardiff.*
Jarvis Dr P. *Department of Botany, St. Machar Drive, Old Aberdeen, Scotland.*
Jewiss Dr O.R. *Grassland Research Institute, Hurley, Nr. Maidenhead, Berks.*
Jones Mr G.E. *Geography Department, University College of Wales, Aberystwyth.*
Jones Mr L. *Grassland Research Institute, Hurley, Nr. Maidenhead, Berks.*
Jones Dr L.H. *Unilever Research Laboratory, Colworth House, Sharnbrook, Beds.*
Jones Mr R. *Agriculture (Crop Husbandry) Department, Institute of Rural Science, Penglais, Aberystwyth.*
Kear Mr B.S. 63 *Lumb Lane, Bramhall, Stockport, Cheshire.*
Keith-Lucas Mr D.M. *Botany School, Downing Street, Cambridge.*
Kennedy Dr J.S. *Entomological Field Station, 34A Storey's Way, Cambridge.*
King Dr J. *Hill Farming Research Organization, 29 Lauder Road, Edinburgh.*
Lambert Mr D.A. *Grassland Research Institute, Hurley, Nr. Maidenhead, Berks.*
Lambert Dr J.M. *Botany Department, The University, Southampton.*
Langton Mr P.H. *c/o The Grammar School, March, Cambs.*
Lee Mr F.A. *Anti-Locust Research Centre, College House, Wrights Lane, London, W.*8.
Lees Mr & Mrs J.C. *Department of Forestry and Natural Resources,* 10 *George Square, Edinburgh* 8.
Lewis Dr M.C. *Department of Botany, The University, Birmingham,* 15.
Lewis Mr R.D. *Department of Zoology, University College, Cardiff.*
Lewis Dr T. *Rothamstead Experimental Station, Harpenden, Herts.*
Leyton Dr L. *Department of Forestry, South Parks Road, Oxford.*
Llewellyn Mr M. *Glasgow University Field Station, Rowardennan, Drymen, Glasgow*
Lloyd Dr P.S. *Department of Botany, The University, Sheffield,* 10.
Lodge Dr E. *Botany Department, Royal Holloway College, Englefield Green, Surrey.*
Long Mr I.F. *Rothamstead Experimental Station, Harpenden, Herts.*
Longden Mr P.C. *National Vegetable Research Station, Wellesbourne, Warwick.*
Longton Dr R.E. *Department of Botany, The University, Birmingham* 15.
Lovett Dr J.V. *Department of Agronomy, University of New England, Australia.*
Luff Dr M.L. *ARC Unit, Close House, Heddon on The Wall, Newcastle upon Tyne* 5.
Lundy Mr H. *Poultry Research Centre, King's Buildings, West Mains Road, Edinburgh* 10.
McCulloch Dr J.S.G. *Hydrological Research Unit, Howbery Park, Wallingford, Berks.*
Macfadyen Prof. A. *The New University of Ulster, Coleraine, N. Ireland.*
Mackay Mr P.J. *Tropical Stored Products Centre, London Road, Slough, Bucks.*
McKelvie Mr A.D. *School of Agriculture,* 41½ *Union Street, Aberdeen.*
MacKerron Mr D.K.L. *Botany Department, University of Aberdeen, St. Machar Drive, Aberdeen.*
McNeill Mr S. *Imperial College Field Station, Silwood Park, Ascot, Berks.*
Machin Dr D. *School of Biology, U.C.N.W., Bangor, N. Wales.*
Maddison Mr P.A. *Imperial College Field Station, Silwood Park, Nr. Ascot, Berks.*
Malcolm Mr D.C. *Department of Forestry & Natural Resources,* 10 *George Square, Edinburgh,* 8.
Marr Dr D.H.A. *Freshwater Fisheries Laboratory, Faskally, Pitlochry, Perthshire.*
Marsh Mr P. *c/o Glenberuie House, The Holt, Farnham, Surrey.*
Martin Dr M.H. *Botany Department, The University, Bristol,* 8.
Mason Mr J.L. 153 *Church Hill Road, East Barnet, Herts.*

Mayhead Mr G.J. *Forestry Department, U.C.N.W., Bangor.*
Meidner Dr H. *Horticulture Research Laboratories, Shinfield Grange, Shinfield, Berks.*
Melican Mr N.J.T. *Unilever Research Laboratory, Sharnbrook, Beds.*
Middleton Mr C.P. *Department of Biological Studies, Lanchester College of Technology, Coventry.*
Milford Dr J.R. *Department of Meteorology, The University, Reading, Berks.*
Millar Mr A. *Monks Wood Experimental Station, Abbots Ripton, Hunts.*
Millar Dr C. *Department of Forestry, University of Aberdeen, St. Machar Drive, Aberdeen.*
Monteith Dr J.L. *Rothamsted Experimental Station, Harpenden, Herts.*
Moore Dr K.G. *Department of Botany, University College of N. Wales, Bangor.*
Morris Mr J.W. *c/o I.R. Martin, 2 Foulser Road, London, S.W.17.*
Morris Dr M.G. *Monks Wood Experimental Station, Abbots Ripton, Huntingdon.*
Morris Mr R.M. *46 Briants Avenue, Caversham, Reading, Berks.*
Motomatsu Mr T. *Rothamsted Experimental Station, Harpenden, Herts.*
Mount Dr L.E. *Institute of Animal Physiology, Babraham, Cambridge.*
Murphy Dr P.W. *School of Agriculture, Sutton Bonington, Loughborough.*
Myerscough Dr P.J. *Department of Botany, King's Buildings, Mayfield Road, Edinburgh, 9.*
New Mr T.R. *Imperial College Field Station, Silwood Park, Ascot, Berks.*
Newbould Dr P.J. *Botany Department, University College, London, W.C.1.*
Newell Dr P. F. *Zoology Department, Westfield College, Kidderpore Avenue, London, N.W.3.*
Newey Dr W.W. *Geography Department, University of Edinburgh, High School Yards, Edinburgh.*
Newman Dr E.I. *Botany Department, University College of Wales, Aberystwyth.*
Newman Mr J.F. *Jealott's Hill Research Station, Bracknell, Berks.*
O'Brien Mr R. *Department of Geography, Queen's College, 17 Bell Street, Dundee.*
Orchard Dr B. *Rothamsted Experimental Station, Harpenden, Herts.*
Østbye Mr E. *Zoological Laboratory, Post box 1050, Blindern, Oslo 3, Norway.*
Packham Mr J.R. *'Dolphins', 58 Dunval Road, Bridgnorth.*
Påhlsson Mr L. *Department of Plant Ecology, Lund University, Ö Vallgaten 17–18, Lund, Sweden.*
Palmer Mrs J. *Rothamsted Experimental Station, Harpenden, Herts.*
Palmer Mr W.H. *32 Fendon Road, Cambridge.*
Pape Mr J. *Department of Agricultural Botany, The University, Reading, Berks.*
Parham Mr M.R. *School of Biological Studies, Wilberforce Road, Norwich, NOR 77H.*
Pascoe Dr H.T. *Portsmouth College of Education, Locksway Road, Portsmouth.*
Pearson Dr M.C. *Botany Department, The University, Nottingham.*
Peckham Dr G.E. *Physics Department, The University, Reading, Berks.*
Pedgley Mr D.E. *Anti-Locust Research Centre, Wrights Lane, London, W.8.*
Perkins Dr D.F. *The Nature Conservancy Research Station, Penrhos Road, Bangor.*
Perrier Dr A. *8 av. du gl. gouraud 78, Viroflay, France.*
Pickrell Mr D.G. *The Dunelm Hotel, Gt. Georges Street, Bristol, 1.*
Piearce Mr T.G. *4 Redwood Avenue, Wollaton, Nottingham.*
Pilbeam Mr J.E. *68 Hough Green, Chester.*
Pollard Mr E. *Monks Wood Experimental Station, Abbots Ripton, Huntingdon.*
Popay Mr A.I. *Botany Department, The University, Manchester, 13.*

Prater Mr A.J. 11 *Newberries Avenue, Radlett, Herts.*
Preece Mr T.F. *Department of Agricultural Botany, The University, Leeds 2.*
Price Jones Dr D. *Jealott's Hill Research Station, Bracknell, Berks.*
Pugh Dr K.B. *Department of Soil Science, The University, Reading, Berks.*
Pullar Dr J.D. *Rowett Research Institute, Bucksburn, Aberdeen.*
Quin Miss F.M. *Botany Department, The University, Reading, Berks.*
Rackham Dr O. *Botany School, Downing Street, Cambridge.*
Rafarel Mr C.R. *Furzebrook Research Station, Wareham, Dorset.*
Rainey Dr R.C. *Anti-Locust Research Centre, College House, Wrights Lane, London, W.8.*
Rains Mr A.B. *Directorate of Overseas Survey, Tolworth Tower, Surbiton, Surrey.*
Ranwell Dr D.S. *Nature Conservancy, Furzebrook, Wareham, Dorset.*
Redfern Miss M. *Department of Biology, Portsmouth College of Technology, Hay Street, Portsmouth.*
Rees Mr T.K. *Department of Biological Sciences, Goldsmiths College, London, S.E.14.*
Reynolds Mr C. *Preston Montford Field Centre, Shrewsbury.*
Reynolds Dr E.R.C. *Department of Forestry, South Parks Road, Oxford.*
Reynolds Miss M.J. *Botany Department, University College of Wales, Aberystwyth.*
Richardson Mr W.D. *Biology Department, Goldsmiths College, London, S.E.14.*
Ricks Mr G.R. *Botany Department, The University, Nottingham.*
Robinson Mr G.A. *Scottish Marine Biological Association, 78 Craighall Road, Edinburgh 6.*
Robinson Mr S.G. *A.R.C. Institute of Animal Physiology, Babraham, Cambridge.*
Rodda Dr J.C. *Hydrological Research Unit, Howbery Park, Wallingford, Berks.*
Rogers Dr J.A. *Hill Farming Research Organization, 29 Lauder Road, Edinburgh 9.*
Rogers Dr S. *Biology Department, Queen Elizabeth College, London, W.8.*
Rogers Dr W.S. *East Malling Research Station, Maidstone, Kent.*
Rorison Dr I.H. *Botany Department, The University, Sheffield, 10.*
Ross Mr M.A. *Department of Agricultural Botany, U.C.N.W., Bangor, Caerns.*
Rowse Mr H.R. *Rothamsted Experimental Station, Harpenden, Herts.*
Russell Prof. E.W. *Department of Soil Science, The University, Reading, Berks.*
Russell Mr R.J. *Glasgow University Field Station, Rowardennan, by Glasgow.*
Rutter Dr A.J. *Botany Department, Imperial College, Prince Consort Road, London, S.W.7.*
Satchell Dr J.E. *Merlewood Research Station, Grange-over-Sands, Lancs.*
Savory Dr B.M., *May & Baker, Ltd., Ongar Research Station, Fyfield Road, Ongar, Essex.*
Schulze Mr E-D., *Institut für Forstbotanik,* 351 *Hann Münden, Germany.*
Seel Dr D.C. *Department of Zoology, University College of Wales, Penglais, Aberystwyth.*
Shackles Mr C.J.D. *School of Agriculture, Sutton Bonington, Loughborough.*
Sheehy Mr J.E. *Welsh Plant Breeding Station, Plas Gogerddan, Nr. Aberystwyth.*
Sheikh Dr K.M. *c/o Imperial College, Botany Department, Prince Consort Road, London, S.W.7.*
Shoun Mr M.J.P. *Rothamsted Experimental Station, Harpenden, Herts.*
Siddorn Mr J.W. *Imperial College Field Station, Silwood Park, Ascot, Berks.*
Sikes Miss S.K. 16 *Cabrera Avenue, Virginia Water, Surrey.*
Sinker Mr C.A. *Preston Montford Field Centre, Nr. Shrewsbury.*

Skellam Mr J.G. 19 *Belgrave Square, London, S.W.*1.
Slater Dr C.H.W. *Pomology Section, East Malling Research Station, Nr, Maidstone, Kent.*
Slater Dr W.G. *Botany Department, South Parks Road, Oxford.*
Smith Dr A. *Grassland Research Institute, Hurley, Berks.*
Smith Dr B.D. *University of Bristol Research Station, Long Ashton, Bristol.*
Smith Dr C.J.S. *Department of Agricultural Botany, The University, Reading, Berks.*
Smith Mr H.G. *Allenby House,* 2 *Gordon Terrace, Edinburgh* 9.
Smith Miss J.G. *Imperial College Field Station, Silwood Park, Ascot, Berks.*
Smith Mr K.G. *MAFF Infestation Control Laboratory, Hook Rise South, Surbiton, Surrey.*
Smith Mr R.T. *Department of Geography, Llandinam Building, Penglais, Aberystwyth.*
Smith Dr S.D. *J. J. Thompson Physical Laboratory, The University, Reading, Berks.*
Smyth Dr J.C. *Paisley College of Technology, Paisley, Renfrewshire.*
Solomon Mr M.E. *Pest Infestation Laboratory, London Road, Slough, Bucks.*
Somerville Mr A. 23 *Blake Street, York.*
Southern Mrs K. *Bureau of Animal Population, Botanic Garden, High Street, Oxford.*
Southorn Miss A. *Botany Department, Royal Holloway College, Englefield Green, Surrey.*
Speight Mr M.C.D. *Institute of Archaeology, Gordon Square, London, W.C.*1.
Spence Dr D.H.N. *Department of Botany, The University, St. Andrews, Fife.*
Spence Mr E.J. *Short and Mason Ltd.,* 280 *Wood Street, Walthamstow, London, E.*17.
Stebbings Mr R.E. *Furzebrook Research Station, Wareham, Dorset.*
Stevenson Dr R.W.H. *Department of Natural Philosophy, The University, Aberdeen.*
Stewart Mr D.H. *Forest Research Station, Alice Holt Lodge, Farnham, Surrey.*
Stewart Dr V.I. *Soil Science Unit, University College of Wales, Penglais, Aberystwyth.*
Stiles Mr W. *The Grassland Research Institute, Hurley, Maidenhead, Berks.*
Stott Mr K.G. *University of Bristol Research Station, Long Ashton, Bristol.*
Stradling Mr D.J. *Department of Agriculture & Forest Zoology, U.C.N.W., Bangor.*
Strangeways Mr I.C. *Hydrological Research Unit, Howbery Park, Wallingford, Berkshire.*
Stringer Mr A. *University of Bristol Research Station, Long Ashton, Bristol.*
Sutton Mr F. *Department of Botany, The University, Sheffield* 10.
Swarbrick Mr J.T. *Botany Department, West of Scotland Agricultural College, Auchincruiue, Ayr.*
Symmons Dr P.M. *Anti-Locust Research Centre, College House, Wrights Lane, London, W.*8.
Szeicz Mr G. *Rothamsted Experimental Station, Harpenden, Herts.*
Taylor Mr J.A. *Geography Department, U.C.W., Penglais, Aberystwyth.*
Taylor Dr K. *Department of Botany, University College, Gower St., London, W.C.*1.
Thomas Mr A.K. *LaMotte,* 2 *Sunnyside, West Lavington, Devizes, Wilts.*
Thomas Mr H. 4 *Hughenden Road, Bristol* 8.
Thomas Miss H.B. *Department of Geography, U.C.W., Penglais, Aberystwyth.*
Thomas Dr J.D. *School of Biological Science, University of Sussex, Falmer, Brighton, Sussex.*
Thomas Mr P.E.L. *c/o Weed Research Organization, Kidlington, Oxford.*
Thompson Mr A.T. *Anti-Locust Research Centre, College House, Wrights Lane, London, W.*8.

Thompson Mr F.B. *Forestry Department, South Parks Road, Oxford.*
Tilbrook Mr P.J. *12 Eliot Place, London, S.E.3.*
Tough Miss A. *Department of Forestry & Natural Resources, 10 George Square, Edinburgh 8.*
Turner Dr D.J. *Weed Research Organization, Kidlington, Oxford.*
Turnock Dr W.J. *Wilterdinkstraat 2, Ede Gelderland, Holland.*
Tyrrell Mr J.G. *Min-y-Mor, South Road, Aberystwyth.*
Úlehla Dr J. *Crop Research Institute, Hrusovany v, Brno, Czechoslovakia.*
Van Emden Dr H.F. *Horticultural Research Laboratories, Shinfield Grange, Shinfield, Berks.*
Varley Prof. G.C. *Department of Entomology, University Museum, Oxford.*
Vlijm Dr L. *Zoology Laboratory, Vrije Universiteit, de Boelelaan 1087, Amsterdam.*
Wadsworth Dr R.M. *Botany Department, The University, Reading, Berks.*
Wagner Miss C.M. *Pest Infestation Laboratory, London Road, Slough, Bucks.*
Waid Dr J.S. *Department of Soil Science, The University, Reading, Berks.*
Wainman Mr H.E. *Agricultural Research Council, London Road, Slough, Bucks.*
Waister Dr P.D. *Scottish Horticultural Research Institute, Invergowrie, Dundee.*
Walker Miss J.M. *University College, Department of Botany, Gower Street, London, W.C.1.*
Walker Mr P.T. *Tropical Pesticides Research Unit, Porton Down, Salisbury.*
Ward Dr L.K. *East Malling Research Station, East Malling, Maidstone, Kent.*
Ward Mr S.D. *Botany Department, University of Aberdeen, St. Machar Drive, Old Aberdeen.*
Wardhaugh Mr K.G. *Anti-Locust Research Centre, College House, Wrights Lane, London, W.8.*
Warley Mr A.P. *School of Agriculture, West Mains Road, Edinburgh, 9.*
Warren Wilson Dr J. *Glasshouse Crops Research Institute, Littlehampton, Sussex.*
Waterhouse Dr F.L. *Department of Natural History, Queen's College, Dundee.*
Waters Mr S.J.P. *Botany Department, Bedford College, Regents Park, London, N.W.1.*
Watson Dr D.J. *Rothamsted Experimental Station, Harpenden, Herts.*
Watts Prof. W.A. *Botany Department, Trinity College, Dublin 2, Ireland.*
Watts Mr W.R. *School of Agriculture, Sutton Bonington, Loughborough.*
Weaving Mr G.S. *N.I.A.E., Wrest Park, Silsoe, Bedford.*
Webb Mr M.J. *Department of Biology, St. Leonard's House, Lancaster.*
Webb Mr N.R. *Department of Zoology, University College, Swansea.*
Welch Dr R.C. *Monks Wood Experimental Station, Abbots Ripton, Huntingdon.*
Wellbank Dr P.J. *Rothamsted Experimental Station, Harpenden, Herts.*
Westlake Mr D.F. *River Laboratory, East Stoke, Wareham, Dorset.*
Wheatley Mr P.E. *Tropical Stored Products Centre, London Road, Slough, Bucks.*
White Mr G.C. *East Malling Research Station, Nr. Maidstone, Kent.*
White Miss R.M. *Department of Forestry and Natural Resources, 10 George Square, Edinburgh 8.*
Wielgolastar Mr F-E., *Botanical Garden, University of Bergen, Bergen, Norway.*
Wightman Mr J. *University of Bristol Research Station, Long Ashton, Bristol.*
Wijmans Dr S.Y. *Jan van Eyckstraat 45bov, Amsterdam, Holland.*
Wilkins Dr D.A. *Department of Botany, University of Birmingham, Birmingham 15.*
Willey Dr R.W. *Department of Agriculture, The University, Reading, Berks.*
Williams Dr J.T. *Department of Biology, Lanchester College of Technology, Coventry.*

Williams Miss M.A. *Department of Microbiology, University College, Cardiff.*
Williams Mr O.B. *c/o Monks Wood Experimental Station, Abbots Ripton, Hunts.*
Wilson Mr J.B. *Sindlesham, 24 Richmond Road, Basingstoke, Hampshire.*
Wilson Miss J.F. *Department of Biology, University of Lancaster, St. Leonardsgate, Lancaster.*
Winter Mr E.J. *National Vegetable Research Station, Wellesbourne, Warwick.*
Woodell Dr S.R.J. *Botany School, South Parks Road, Oxford.*
Woolhouse Mr A.R. *Department of Biology, University of Lancaster, St. Leonards Gate Lancaster.*
Yeates Miss M. *Anti-Locust Research Centre, Wrights Lane, London, W.8.*
Young Mr C.W.T. *Alice Holt Lodge, Forestry Commission Research Station, Farnham, Surrey.*

AUTHOR INDEX

Those page numbers given in ordinary type indicate where the author's work is quoted in the text. The page numbers in **bold type** refer to the page where the full reference of the paper quoted is given.

SUBJECT INDEX